Studies in Human Ecology and Adaptation

Volume 12

This series reflects a growing trend in the social sciences towards interdisciplinary study and research spanning a large number of disciplines, such as archaeology, anthropology, environmental sociology, geography and demography. It focuses on the adaptation of humans to their environment throughout the world. Within the environmental sciences, the field of human ecology is the study of the ways in which human social behavior is affected by environmental factors and events, including but not limited to natural resources, anthropogenic environmental changes and the interactions of competing or co-operating human groups. It encompasses a broad perspective that views the biological, environmental, demographic, and technological aspects of human existence as interrelated. Because the approach is multidisciplinary, it brings new and often unexpected insights to many topical issues. This series publishes cutting-edge and critical research in the field of human adaptation to their environments. Both edited volumes and monographs are welcome.

Michael Rosenberg

The Dynamics of Cultural Evolution

The Central Role of Purposive Behaviors

Springer

Michael Rosenberg
Philadelphia, PA, USA

ISSN 1574-0501
Studies in Human Ecology and Adaptation
ISBN 978-3-031-04862-3   ISBN 978-3-031-04863-0 (eBook)
https://doi.org/10.1007/978-3-031-04863-0

This Springer imprint is published by the registered company Springer Nature Switzerland AG
The registered company address is: Gewerbestrasse 11, 6330 Cham, Switzerland

Preface

This book is an exploration of the structural complexity inherent in the thing we call culture, as well as of how that complexity feeds into the dynamics of cultural evolution. Its roots lie in a 1994 article I wrote whose primary aim was to argue that, while culture change can proceed via both microevolutionary (i.e., selectionist) and macroevolutionary (i.e., punctuational) processes, new cultural systems always come about through a punctuational episode. To do that required me to first make the case that cultural systems are entities bounded in time as well as space, and to do that I applied Niles Eldredge's concept of biological evolution proceeding through the interaction of two distinct hierarchies of entities. The first consists of entities that reproduce and evolve by means of differential reproduction, and the second consists of entities that interact economically and through those interactions provide the raw material for differential reproduction. However, my focus was on entities and the subject of hierarchies was to me at the time merely a stepping stone to dealing with them. The result was that I did not do their complexity justice. It eventually became clear to me that the system of hierarchies I proposed in that article was inadequate for encompassing the complexity of what we commonly call culture because I had omitted explicit reference to behavioral systems as distinct systems. Specifically, in 2003 Prentiss and Chatters published an article in *Current Anthropology* that treated behavioral systems – adaptations – as distinct entities, which had their own Baupläne. That prompted me to try and reconcile their understandings of Bauplan with mine, which had largely focused on culture as information systems and skirted the issue of what specifically to do with behavioral systems. Initial attempts to deal with the entirety of it all took the form of draft articles, but these were either too long to be acceptable as articles or, if shorter, too scattershot and sketchy. That left no choice but to try and pull it all together in the form of the first seven chapters of this book. That, in turn, allowed me to conclude the project by coming full circle and revisiting other published articles in order to situate the ideas contained in them in the context of the model of cultural dynamics being proposed.

Along the way many friends, colleagues, and anonymous reviewers influenced my thinking by commenting on the aforementioned draft articles that went nowhere, the published articles, or in the course of conversations, with several particularly memorable conversations taking place among the participants of a symposium on macroevolutionary processes held at the 70th Annual Meeting of the Society for American Archaeology in 2005 and organized by Anna M. Prentiss, James C. Chatters, and Ian Kuijt. So, without further ado and in no meaningful order, I would particularly like to thank Anna Prentiss, James Chatters, Ian Kuijt, Charles Spencer, Niles Eldredge, Robert Bettinger, Stuart Fidel, David Anthony, Dan Varisco, Brian Peasnall, Mindy Zeder, Tom Rocek, Brian Hayden, Lee Lyman, Michael O'Brien, Peter Richerson, Ward Goodenough, and several anonymous reviewers for sharing their detailed thoughts with me. While not all of them necessarily agreed with my views on the subject being discussed, they did contribute to my thinking on the subject and I appreciate the time they devoted to offering suggestions as well as constructive criticism that prompted me to start thinking in new directions. I am certain that were it not for them, my thinking would have stagnated and this book would never have come together however well it did.

On a more material level, I would also like to thank the staff of Springer Nature for shepherding me through the publication process. In particular, I would like to thank Theresa Krauss and Christi Jongepier-Lue for seeing value in my project and their help in getting my manuscript in front of the correct eyes within Springer; Shinjini Chatterjee for managing the acquisition and publication process, as well as for her very helpful suggestions along the way, including the suggestion that I considered it for inclusion in the *Studies in Human Ecology and Adaptation* series; Daniel G. Bates and Ludomir R. Lozny, the series' editors, for deeming it worth including in their series; and Daniel Jagadisan, the project coordinator, and Kali Gayathri for their editorial and administrative help.

Philadelphia, USA Michael Rosenberg

Contents

About the Author

Michael Rosenberg is a retired Professor of Anthropology at the University of Delaware, Wilmington, USA. His primary research interests are the mechanics of cultural evolution and understanding the beginnings of food production, settled village life, and social complexity in the context of those mechanics. In the late 1980s-early 1990s, he participated in the Tigris-Euphrates Archaeological Reconnaissance Project in Anatolia, Turkey, and in the 1990s, he discovered and directed the salvage excavations at the early Neolithic sites of Hallan Çemi and Demirköy in southeastern Turkey. He has published over 40 journal articles in reputed journals and several book chapters in edited volumes with international publishers.

Chapter 1
Introduction

> *"Evolution is not a cause, but the description of a process."*
> *George Gabriel Stokes*

Dunnell once correctly observed that "there can be no profitable discussion of evolutionary theory . . . unless there is initial agreement on precisely what evolution means," and he went on to succinctly define it as "a particular framework for explaining change as [the] differential persistence of variability" (1980: 38). The problem with this nutshell definition (on which he later elaborated) is not its succinctness; it is that, while true enough so far as it goes, it proceeds as if it is the only correct definition of the word "evolution." However, it is arguably just the only correct definition for cultural evolution, just as it is the only correct definition for biological evolution. But there is another accepted definition of evolution, one properly associated with Herbert Spencer rather than Charles Darwin, which is still legitimately used by, for example, astronomers and is in fact closer to the meaning of the Latin word *evolutio* (unrolling, unfolding) from which the term evolution is derived.

In this second meaning, evolution refers to a set progression along a path predetermined by starting conditions and leading inexorably to an ultimate result (e.g., see Gould, 2002b: 248ff). Thus, any G2V type main sequence star such as our sun will ultimately evolve into a red giant, and later a white dwarf, and still later a black dwarf; and, it will inevitably do so every time. This second, Spencerian view of evolution had dominated Anthropology from its very origins in the thinking of the unilinear evolutionists until the last quarter of the twentieth century and was what Dunnell was rightly criticizing. But they were not relying on a blatantly invalid theory rooted in a lack of understanding as to what evolution is; they were relying on an application of the wrong theory to the evolution of culture, one that nevertheless works to explain other phenomena rather well, and superficially seemed applicable if one starts with a progressive view of cultural evolution and the human condition and applies that to the patterns in prehistory.

That is, the sometimes contingent nature of cultural evolution—e.g., you cannot get urban societies without first having food production; you cannot have extra-familial political hierarchy without first abandoning egalitarianism—superficially

M. Rosenberg, *The Dynamics of Cultural Evolution*, Studies in Human Ecology and Adaptation 12, https://doi.org/10.1007/978-3-031-04863-0_1

resembles such a process of progressive "unfolding"; and such contingent transformations independently repeated themselves often enough in the past to *seemingly* indicate that cultural evolution was such a process of "unfolding." But the natural and social worlds, unlike the physical world, are not governed by laws, such as those that govern physics and chemistry and which make it possible both to have and to predict such set "unfoldings." This makes the biological theory of evolution, one that relies on variability in potential outcomes, the more appropriate one when dealing with the evolution of culture, despite its often contingent nature. The question then becomes how one can reconcile the variability of potential cultural outcomes with the frequency of parallelisms and convergences in actual cultural evolutionary outcomes. The answer is that human beings have behavioral tendencies, as studied by ethologists and psychologists, which bias the likelihood of some potential outcomes over others but do not determine them by any stretch of the imagination.

This Spencerian, progressive view of evolution consciously or unconsciously still influences the thinking of many to one degree or another in yet another way, producing a distinct teleological bias to questions concerning the evolution of culture. That is, the present state of whatever cultural phenomenon is being investigated implicitly becomes the *telos*, the primary reference point from which the past is viewed, with questions being phrased in terms of how/why/when did the present state evolve instead of how/why/when did it evolve when it did *and not evolve in other potential contexts*. From the progressive perspective, more often than not, the cases of "why not" were and still are skirted for being self-evident evolutionary dead ends, in that they did not continue on the path to the present state, which is considered inherently superior to such alternatives. Such "failures" could perhaps be because culture was simply not ready or environmental conditions were perhaps not conducive, but in any case, such dead ends are of no interest because they do not explain the superior outcome of present interest. Even from a properly structured evolutionary perspective, the question of "why not in other contexts" also tends to be skirted on the implicitly self-evident grounds that it was just not selected for, either because the proper selective forces were not in operation or the relevant innovation did not appear to be selected for, with little consideration given to why not in the case of the latter.

They are thus, from both perspectives, merely relatively unimportant side branches on the tree whose apex is the present state; they are the cultural equivalent of Neanderthals, with the only significant difference from the Neanderthal analogy being that Neanderthals got and still get appreciably more attention from researchers interested in the evolution of the genus *Homo* than do cultural "dead ends" at the hands of researchers interested in the evolution of culture. The point is that just as the study of Neanderthals can inform the evolution of our species, so can the study of cultural side branches. As Gould (e.g., 1976) was fond of pointing out, biological evolution produces a much bushier set of outcomes than our present position looking back down our specific branch and from there down the trunk of the tree tends to give it credit for and one has only to look around the world to see that cultural evolution is just as "bushy" if not more so. That bushiness requires addressing, and that in turn requires us to first take a closer look at just what the "proper" biological theory of evolution involves.

While discussions of evolutionary theory in Anthropology and Archaeology have tended to focus on the primacy/relevance/issue of progressive versus Darwinian evolution, until the beginning of this century, the debate has largely proceeded as if the structure of Darwinian evolution is a settled matter. However, the biological theory of evolution is not static, nor is it a monolithic, universally agreed-upon theory within biological circles. Rather, it is an evolving synthesis that only began with the "Modern Synthesis" that took form around the middle of the twentieth century. Some of the synthesis' constituent theories, such as genetics, had previously been offered as alternatives to Darwinian evolution, and such theoretical competition did not cease with the formulation of the Modern Synthesis. Some more recent theories, such as the theory that evolution proceeds by a series of punctuated equilibria (Eldredge & Gould, 1972; Gould & Eldredge, 1977), which was initially offered as an alternative to Darwinian gradualism *as regards the appearance of new species*, and the issue of group selection (e.g., Sober & Wilson, 1998), which is a debate that goes back to Darwin himself, have yet to fully find their place within this evolving extended synthesis, nor is there yet agreement on whether the proper approach should be reductionist and focus on the gene, or taxic and focus on higher-order entities (e.g., see Sterelny, 2001).

The point is that Archaeology, which we can, for working purposes, loosely define as the study of the undocumented aspects of the cultural past through the remains of material culture, and also Anthropology in general, to the degree that it is also concerned with that past as it relates to the evolution of culture, rely heavily on theories borrowed from other disciplines, and no single discipline more so than Biology. But the fact that a given borrowed biological approach to the data is productive (or at least seems to be so) does not alter the fact that there are at any given time other biological approaches, which may also be as productive, if not more so, that have not yet all achieved the same level of popularity in their new disciplinary home. This is not because the biological evidence refutes them, but due more to a lack of sufficient discussion. Nor are they necessarily in contradiction to some more established elements. Thus, elements of biological theory that are not yet widely applied to the evolution of culture are worth considering in order to evaluate their explanatory power regarding the evolution of culture.

While Dunnell's observation, quoted earlier, concerning the importance of understanding exactly what evolution means is certainly true, a clear understanding of the entities said to evolve is a prerequisite for dealing with the issue of what evolution means, because it is axiomatic that what is claimed to evolve bears on the issue of how it evolves. As aptly noted by Mayr (1942: 114) with respect to biological evolution, ". . . the question of 'how do species originate?' cannot be discussed until we have formed some idea as to what a species is."

However, even in Biology, the issue of entities is nowhere near as cut and dried as it sometimes appears to be to observers outside the circle of active discussants. For example, in Biology, before we can even begin talking about whether the proper evolutionary locus is the gene (e.g., Dawkins, 1976) or some higher-order entity (e.g., Gould, 2002a), it is necessary to first address the issue of what such larger entities are. The issue of entities is even murkier in Anthropology and Archaeology.

Anthropology has historically had great difficulty in dealing with the issue of cultural entities (e.g., see Whiting, 1964; Brumann, 1999). Thus, while attempts to define the concept of culture abound in the anthropological literature (e.g., see Kroeber & Kluckhohn, 1952), the opposite is the case with the categorically messier task of delineating individual types of cultural entities, with their seemingly endless and inconsistent expression in sub- and macro-sets, not to mention their vague vertical and horizontal boundaries (but see Goodenough, 1981, 1999; Boyd et al., 1997; Bashkow, 2004). Archaeology has historically shared that same difficulty (e.g., see Clarke, 1968), despite its heavy analytical reliance on the use of spatio-temporal culturally defined categories larger than the individual artifact.

The first point here is that significantly more attention has thus far been paid to the proposed proper mechanics of cultural evolution, or to detailing specific cultural adaptations and adaptive behaviors, than has been paid to gaining an understanding of exactly what kind of entities there are that are said to evolve by either actively changing (i.e., adapting) or becoming changed by selective forces (i.e., becoming adapted). While this avoidance is understandable in those reductionist quarters that do not generally hold for the existence of evolutionary entities beyond the individual culture bearer's cultural proxy in the form of, for example, an artifact (or even that proxy's individual traits), grappling with the 'entity issue' is an unavoidable necessity for those quarters that do hold for the existence of higher-order entities, or at least logically should because they tend to view culture systemically. The second and more important point is that Biology has spent considerably more effort on the issue of entities than has Anthropology, and an examination of how Biology has grappled with the species concept arguably illuminates a path through the quagmire of cultural entities.

All this is further muddled by a terminological problem that tends to lump diverse culturally related entities under the same general rubric of culture. Thus, culture is often used interchangeably to refer to an information system by those who view culture as an ideational system, a behavioral system by those who view culture in materialistic terms, and a social group that is said to share either or both of the above two. Leaving aside momentarily the long running debates concerning the relative primacy of ideas versus behaviors in the matter of culture change and which of those two is the proper view of culture, the third usage actually refers to something only peripherally related to culture: a society. As has already been noted by others (e.g., Geertz, 1973), societies, while structured by culture, are first and foremost social groups made up of biological individuals and thus not the same as culture, which is not biological in nature (e.g., see Kroeber, 1917). Thus, it can be argued that while the evolution of the culture (whatever that is claimed to mean) a given social group shares impacts the somatic and reproductive well-being of the individuals that constitute that social group, the entities that constitute the group, as well as the group itself, are not subject to cultural evolution and are instead subject to biological evolution, albeit with reference to the cultural systems they carry and the directions those systems evolve in.

In other words, cultural entities are not identical to biological entities, and allowance must be made for those differences in any discussion of entities and the processes by which they evolve. To begin with, while human beings are organisms and as such subject to biological evolutionary processes, they are also carriers of culture, which is not "organic" (i.e., alive) either in whole or in part, even though it structures the organization of human groups, whose memberships are. As noted by others, both have/are coevolved (e.g., see Durham, 1979, 1991), and thus, any discussion of entities and evolution must address the nature of that relationship. But the evolutionary processes that govern cultural entities are not necessarily identical to the ones that govern the human entities that are structured by them. As noted by Kroeber (1917) a century ago and most thoroughly explored by Boyd and Richerson (1985), the transmission mechanisms (i.e., learning) for cultural information are, by their very nature, considerably more variable than those for the transmission of genetic information (i.e., inheritance), giving cultural evolution a decidedly Lamarckian character (in addition to its Darwinian elements) not present in biological evolution. Add to that the fact that the selective forces operating on cultural variability are different, primarily due to the addition of conscious or unconscious decision-making at work in the form of what is usually referred to as "cultural selection," and you are confronted with a considerably more complex process than that which operates to produce biological evolution.

Furthermore, there is also the related issue of agency's role in proximate causation. Evolutionary approaches to the study of culture have proceeded much as do biological approaches to evolution and treat the entities in question as for all practical explanatory purposes the passive objects of selective forces—selection being ultimate causation—without explicitly acknowledging that they might also be reacting purposively to those same forces. This approach in Biology has recently been most comprehensively criticized by Walsh (2015), and one can make an even stronger case for such implicitly considered passivity in the face of environmental conditions being in error as regards cultural evolution.

While genes are the ultimate source of biological novelties, innovation is the ultimate source of cultural novelties. More to the point, notwithstanding Dawkins' (1976, 1989) metaphorical use of "selfish" in reference to genes, genes are obviously incapable of acting, much less acting selfishly or even self-interestedly to produce potentially beneficial novelties, while organisms are so capable of self-interestedly producing potentially beneficial cultural novelties. And none are more so capable than humans given their substantially greater intellect than other organisms and the concomitantly greater ability to however accurately or inaccurately, as the case may be, relate cause and potential effect. This adds yet another Lamarckian element to cultural evolution in the form of biased proximate causation.

With respect to the evolution of cultures, while the progressive perspective implicitly views proximate causation in the form of such motivated innovation as inherently adaptive/superior, in that cause and effect are treated as if they are always correctly perceived and the resultant innovation always beneficially functional, the dominant biological perspective currently applied to cultural evolution tends to explicitly or implicitly view proximate causation as inherently unknowable scientifically. Both positions are arguably incorrect for being extreme. Human behavior is neither nomothetically patterned nor motivationally random. It is, however, subject to tendencies, which bias outcomes but do not determine them, and those behavioral tendencies are scientifically knowable within the framework of Psychology and Ethology, if not Archaeology or Anthropology. We borrow liberally from Biology. Why cannot we also appropriately borrow from Psychology and Ethology?

More importantly and in an entirely different vein, virtually all perspectives tend to view culture as completely open-ended and as capable of evolving in any one direction as in any other direction short of any environmental constraints. Both are arguably wrong on this count as well, and this is arguably the more serious error.

The problem with viewing culture as completely open-ended is implicit in the classic definition of culture still found in the overwhelming majority of textbooks. It begins with the words: "Culture ... is that complex whole. ..." (Tylor, 1871). And while no one would nowadays seriously argue that a culture is ever a *fully* integrated systemic whole, one would have an equally difficult time arguing that it is simply a disjointed aggregation of elements and that it does not tend dynamically toward integration, with elements constantly changing and others secondarily, tertiarily, quatinarily, etc., changing to accommodate the changes that initiated their changes. In other words, culture is in a sense much like a metaphorical inchworm, in perpetual motion in one direction or another, with its tail end always trying to catch up to its front end, but never quite making it because the front end is always moving ahead. This lack of total integration notwithstanding, in dealing with culture, we are dealing with a *system*, and systems are both structurally integrated and constrain potential changes to their existing structure simply by virtue of having a structure to begin with. In biological theory, this principle is enshrined in the dictum that the vast majority of mutations, by virtue of being random, are ultimately selected against. That is, the odds of a random change actually improving a functioning organism—a functioning biological *system*—are very slim indeed. This is simply because most random changes cannot be integrated into the existing system's structure in a manner that maintains the system's functional integrity and viability, much less improve on it.

In the case of biological organisms, the organism itself has no say in what biological novelties it carries, having been born with its specific complement of traits regardless of whether any specific novelty hampers the efficacious operation of the system that is the organism itself (as a complex whole). In the biological domain, selective forces tend to remove such hampered systems from the competition, and the more hampered they are, the more quickly they tend to be removed. Cultural novelties, more commonly referred to as innovations, are not subject to that same kind of deterministic expression due to the Lamarckian elements inherent in cultural evolution. The mere fact that someone has an idea for a different way to do something does not automatically mean that others will incorporate it into their propriospect, and the more it clashes with other elements of that propriospect, the less likely they are to do so. Likewise, the mere fact that someone engages in a behavior that deviates from existing behavioral norms does not automatically mean that others will emulate it. The more it interferes with other behaviors geared to achieving desired goals, the less likely it is to be emulated, and the more likely the individual so behaving will be considered "odd" or worse and treated accordingly. In other words, potential cultural novelties can be excluded from ever becoming part of the system if they are viewed to sufficiently conflict with the system as a whole. Thus, no cultural system is completely open-ended. Rather, it is truly open-ended only in those directions left open by the existing structure of the system while at the same time being substantially closed to change in other directions.

The ultimate problem is that the study of evolutionary culture theory, Durham's (1990) umbrella term for the larger body of theory pertaining to cultural evolution, is currently fragmented without an overarching theory that unifies evolutionary theory and culture theory as a whole. That is, the emphasis to this point has been on those aspects of culture theory pertaining to cultural transmission and cultural selection and their roles in the mechanics of cultural evolution and to the virtual exclusion of those aspects of culture theory dealing with just what it is that is said to evolve; and it is those latter aspects of culture theory that suggest that there is more to the mechanics of cultural evolution than has thus far been examined.

That emphasis on cultural transmission and cultural selection was justified, given the progressivistic perspective that had dominated previously. It was also understandable, given the glaring need for cultural counterparts to the roles that inheritance and natural selection play in biological evolution, if one wanted to loosely model cultural evolution on biological evolution. Unfortunately, the exploration of culture theory as it potentially relates to cultural evolution more or less stopped there. I would argue that this stems at least in part from the reductionism that dominated biological theory (e.g., Dawkins, 1976, 1982, 1986) at the time of its importation into the social sciences in the last quarter of the twentieth century and which is arguably still the dominant view in its new home as well—a cultural version of founder's effect. That is, with few exceptions, the issue of entities—what is said to evolve—focused largely on the individual cultural trait, artifact, behavior, or individual. That initially narrow focus has gradually expanded to an exploration of higher-order entities such as societies and cultures, with the highly beneficial side effect that what were initially disparate approaches to the evolution of culture have begun to find common ground. However, a thorough exploration of just what evolves and the broader mechanics of how they potentially do so remains to be done.

That exploration is what follows in this book. It begins with a review of the extended synthesis in Biology, what elements of it have been borrowed into the social sciences, and what other elements arguably should. The purpose is to lay the groundwork for addressing the fundamental issue of cultural entities, whether they can be said to exist as discreet entities, and if so, their various types, characteristics, and how they can be defined, as well as the kinds of selective forces working on each type of entity. The system employed is modeled on Eldredge's (1985a) biological distinction between reproductive and economic hierarchies of entities but modified to accommodate the greater number of culturally structured entities than there are of biological entities.

For something to be a discreet entity, as opposed to an arbitrarily defined segment of some continuum—an ephemeral class—it must be bounded with respect to other entities both diachronically and synchronically. It must have a "birth," however fuzzy, at which it comes into being, and a "death," at which it ceases to exist, as well as an external boundary, again however fuzzy, in the time between that birth and death, which separates it from other entities of like or different kind. Within those borders, it does not have to be unchanging, but it does have to be unchanged with respect to its basic defining structural characteristics—its cultural "morphotype" so to speak. Were it not, there could be no such thing as its birth or death.

The point of doing all this is not by any stretch of the imagination to construct some kind of cultural taxonomic system for the purpose of categorizing entities. The variability inherent within each type would render such an exercise pointlessly sterile more often than not and usually not worth the effort required for doing so. It is simply to construct a heuristic device in order to demonstrate that such various kinds of entities can, first of all, be said to exist, that there is a systematic order to their existence, to show how they are integral to a complex of dynamic evolutionary processes, to illustrate the full range of evolutionary dynamics that drive the evolution of culture in terms of both micro and macro evolution, and to show how agential actions are stimulated by both cultural and environmental inputs in the context of those evolutionary dynamics.

Lastly, evolution is a product of chance, yet there is a high degree of parallel and convergent evolution evident in the evolution of cultural systems, not all examples of which are ascribable to simply diffusion. This strongly suggests that proximate causation in the evolution of cultural systems is not a product of only random chance. Rather, humans, like other animal species, are imbued with behavioral tendencies that bias the form that purposive actions tend to take in response to given environmental stimuli, meaning that proximate causation in the form of agential behaviors can be knowable by means of Psychology and Ethology. The exploration concludes by proposing that these tendencies played out fairly consistently to drive the evolution of the two most significant evolutionary transitions that serve as the basis for the human cultural world as it presently exists: the evolution of food-producing economic systems and the evolution of sociopolitical systems built around directive leadership.

Chapter 2
The Evolution of Evolutionary Theory in Biology

> *"The question of 'how do species originate?' cannot be discussed until we have formed some idea as to what a species is."*
> *Ernst Mayr*

The Evolving Extended Synthesis and Cultural Evolution

Scientific bodies of theory, just like other aspects of culture, can be said to evolve, in that they are characterized by variability and selection (see Hull (1988) for an interesting, albeit arguably flawed, attempt to systematize this process and Gould (2002a: 543) for an interesting observation concerning the impact of intellectual 'founder's effect' on the evolution of scientific theories, one that arguably applies to the importation of evolutionary theory into the social sciences). Ideally science allows a variety of ideas to freely compete for acceptance, with the best adapted (to the then-available data environment) tending to ultimately gain acceptance at the expense of less well 'adapted' ideas. Evolutionary theory in Biology is no exception, having evolved from its Darwinian roots into the 'Modern Synthesis' of the mid-twentieth century and that neo-Darwinian synthesis' still-more modern manifestations as an evolving whole, usually referred to nowadays as the extended synthesis, characterized by a continuing intellectual melee between often competing ideas as to how evolution proceeds. The point is that evolutionary theory has never been monolithic and has always been characterized by competing ideas that sometimes could be judiciously reconciled and sometimes not. The crafting of the Modern Synthesis in the mid-twentieth century was certainly not the end of this process. It simply relegated to oblivion in the minds of an overwhelming number of scholars the purely genetic theories of evolution (e.g., De Vries, 1905, 1909, 1910; Goldschmidt, 1940) that had been offered as alternatives to Darwinian evolution subsequent to the rediscovery of the data pertaining to Mendel's experiments. It did so by reconciling genetics and Darwinian evolutionary mechanics into an acceptably coherent whole in the context of the then-available data. However, other debates, such as those pertaining to the units of selection, group selection, and the definition of species, continued, and new ones inevitably arose.

M. Rosenberg, *The Dynamics of Cultural Evolution*, Studies in Human Ecology and Adaptation 12, https://doi.org/10.1007/978-3-031-04863-0_2

Five of these debated points are particularly pertinent to any discussion of cultural evolution. All five directly or indirectly revolve around the issue of entities. The first of these is the species concept. This is because the issue of how to define species as entities, and whether such entities can even be said to exist, bears on the issue of how to define cultural entities and whether such cultural entities can also be said to even exist. As aptly noted by Mayr (1942: 114), "[t]he question of 'how do species originate?' cannot be discussed until we have formed some idea as to what a species is," and the same can be said concerning cultural systems. The second issue, also implicit in Mayr's quote above, is related to the first. It is the question of how species, if they can in fact be considered evolutionary entities, come into being and how they cease to exist, because an entity must be bounded in time as well as space. This bears on the issue of how any potential cultural entities can be said to come into being and subsequently cease to exist. The third is whether genes or organisms are central to evolution. This bears on the issue of agency in evolution because organisms can act purposively while genes cannot. The fourth is the question of whether selective forces can legitimately be said to operate on any higher-order biological entities above the organism, such as species. This is because if multilevel selection can be properly said to operate at all, it must perforce operate on any hierarchy of entities that are claimed to evolve, be they biological or cultural. Lastly, there is the issue of organismic agency's role in biological evolution. This is because humans are arguably the most complexly agential organisms on the planet and behavior is integral to cultural transmission.

The Species Concept: Adaptation Versus Reproduction

Evolution, by definition, involves descent with modification. Thus, in order for there to be descent, the entities that evolve must have the capacity for "more-making"—the ability to make more, not necessarily of themselves, but of like kind (Eldredge, 1985a). Adaptation, on the other hand, pertains to interactions between entities and their environments, which environments include other entities of both like and different kind. Ecological adaptations are essentially economic in nature in that they are involved with energy conversion or energy transfers. Reproductive adaptations, on the other hand, are involved with 'more-making' (see Hull, 1980: 318). The point is that adaptation is not evolution, but a product of evolution, with the economic interactions themselves being one of the driving forces of evolution, in the form of natural selection, and reproductive interactions being another, in the form of sexual selection (see Eldredge, 1985a). The significance of that distinction between evolution and adaptation in the biological (and in the cultural) domain is that evolutionary entities are usually not the same as economic ones. Hull (1976, 1980) referred to the former as replicators, the latter as interactors, and defined selection as the "process [by] which the differential extinction and proliferation of interactors cause the differential perpetuation of the replicators that produce them" (1980: 318). Hull made that distinction specifically to further a reductionist view of biological evolution wherein organisms are merely interactors and it is genes that are the primary if not sole true replicators and hence units of evolution (e.g., 1994, 2001: 48). While Hull's ultimate restriction of replicators to genes is a crucial point of contention that we will return to, the distinction he makes between interactors and replicators is both correct and important.

As noted by Eldredge (1985a), the species concept was embedded in the Modern Synthesis (e.g., Mayr, 1942; Dobzhansky, 1951)—much like the conception of individual cultures is *still* embedded within Anthropology—with a variable, context-dependent ontology that straddled this distinction between interactors and replicators without formally acknowledging that the distinction exists. That is, when viewed over time, species seemed to be nothing more than "arbitrarily delineated segments of evolving continua, defined and recognized by some convenient combinations of characters gradationally defined" (Eldredge, 1985a: 90). However, when viewed spatially at any specific point of time, they could be seen as discreet entities that are spatially bounded with respect to other species. Cultures tend to be viewed the same way.

The "arbitrarily delineated" conception of species stemmed from a focus on economic adaptation, natural selection, and a species' constituent individuals as interactors. Such a focus fosters a class-like view of species as variable collections of phenotypically similar organisms, anatomically tracking shifting adaptive peaks over time due to natural selection, that are almost *incidentally* restricted to (successful) reproduction with conspecifics by virtue of the genetic compatibility underlying the common adaptations shared by constituent members. In contrast, the "bounded" conception of species stemmed from a focus on species as reproductive communities, geographically, or even just reproductively isolated from other such plexuses. Isolation is not specifically due to genetic incompatibility, but —because adaptations can be reproductive as well as economic (e.g., see Eldredge & Cracraft, 1980)—by a somewhat distinctive set of phenotypic attributes that make conspecifics recognizable potential mating partners and others not (see Paterson, 1985).

> To see species as clusters of similarly adapted organisms is to see them ... with no 'bridgeless gaps' between them and their nearest phylogenetic neighbors.... [On the other hand], to see species ... primarily as reproductive communities ... isolated from other such groups by means of specific mate recognition systems (SMRSs) ... amounts to saying that species are stable entities. (Eldredge, 1985a: 157–158)

In other words, to view species ecologically/economically is to see them as ephemeral classes, fostering reductionism—"mere names of convenience attached to segments of evolving continua without clear borders" (Gould, 2002a: 604). The problem with this view is that "no ecologist has [yet] adduced a theoretical structure that sees entire species [monolithically] integrated into some ecological system" (Eldredge, 1985a: 159). For example, as noted by Eldredge, it is impossible to speak of *the* adaptation of the species *Panthera pardus* (leopards), with its huge range (Africa to Indonesia) and diverse prey within that range. To that, I would add another obvious example, *Orcinus orca* (orcas), for whom the same applies, given their global range and diverse prey species that range from herring in the North Sea to southern elephant seals in the Southern Ocean. Moreover, with just a modicum of effort, one can easily come up with dozens if not hundreds more such examples. The point is that "in a purely ecological, economic, energetics sense, species do not exist," though "a species can play an economic role [in an ecological system] in the additive sense of its component populations and ... organisms" (Eldredge, 1985a: 159).

For that reason, as detailed by Eldredge (1985a, 1999, 2015), it is the reproductive aspect of the species concept, as most strongly advocated by Mayr (1942), that came to dominate the Neo-Darwinian synthesis; and to see species reproductively is to see them as definable individual entities. However, the problem with even this view of species remained that while spatially bounded (reproductively) at any single point in time and continually so bounded at successive points in time, species were not similarly seen to be bounded temporally (i.e., initially and terminally) in the Modern Synthesis. Thus, in evolutionary terms, they remained temporal ephemera—'mere names of convenience' for a synchronic snapshot of an evolving continuum, paving the way for the evolutionary reductionism of gene selectionism to dominate the neo-Darwinian synthesis (e.g., Williams, 1966; Dawkins, 1976, 1982, 1986). That is, a reductionist would argue that the ecological nonexistence of species, coupled with their temporal nonexistence, is precisely the point; species are not discreet entities in any way, shape, or form and hence irrelevant to the process of biological evolution (e.g., see Dawkins, 1986).

However, leaping from the fallacious, ecological view of species and the temporal gradualism of the Modern Synthesis to the conclusion that species do not in any other way exist as definable entities ignores the very real spatial boundaries of species, which were never seriously in question. Simply put, nature is not some ecological "great chain" of *individual* beings. Spatially diverse populations of leopards (*Panthera pardus*) may be characterized by distinct behavioral economic adaptations and even physical differences, but leopards do not spatially grade into lions (*Panthera leo*) or tigers (*Panthera tigris*); there is a fairly large gap between the point where being a leopard stops and being a lion or tiger starts.

As noted above, the crux of the 'entity problem' with a reproductive view of species was the absence of any temporal boundaries in the phyletic gradualism of the mid-century neo-Darwinian Modern Synthesis. Definable entities have temporal boundaries, however fuzzy, as well as spatial boundaries. They have beginnings and ends—births and deaths—as well as external borders. The extinction of a given species constitutes its end, but its beginning remained highly problematic in the slow, incremental phyletic gradualism of the neo-Darwinian synthesis. However, the full set of temporal bounds necessary for a fully developed view of species as *historic* (i.e., evolutionary) individuals came with the realization of the importance and prevalence of stasis in evolutionary history as visible in the paleontological record (Eldredge & Gould, 1972; Gould & Eldredge, 1977).

Beginnings and Endings: Stasis and Punctuated Equilibria

The neo-Darwinian synthesis viewed the evolution of species in classic Darwinian terms—evident in the critical importance of the vastness of time to Darwin—as the *slow and incremental* (i.e., gradual) accumulation of change over time. Given enough time, two or more populations isolated from each other would gradually diverge sufficiently as to eventually warrant (name-of-convenience) designation as separate species, or enough change would accumulate within a single lineage as to warrant a separate (name-of-convenience) designation from the ancestral stock. However, while such a view was leading to the tantamount abandonment of the species concept in some biological quarters, in Paleontology, the very slow and incremental nature of phyletic gradualism was coming to be challenged by a fossil record that contradicted gradualism and the development of a theory that proposed that, based on the actual fossil record, speciation was not a very *steadily incremental* process at all. Rather, it was proposed that fossil species did not change much over the course of their evolutionary existence and that new species came into existence very rapidly, thus 'punctuating' the evolutionary 'equilibria' that marked the long intervening periods of evolutionary stasis.

This theory of evolution proceeding as a series of punctuated equilibria (Eldredge, 1971; Eldredge & Gould, 1972; Gould & Eldredge, 1977; also see Stanley, 1979, 1981), drawing partly on Mayr (e.g., 1942, 1954, 1963) and Simpson (e.g., 1944, 1953), departed from the neo-Darwinian synthesis in one profoundly important respect. It rejected the proposition that speciation results simply from the slow and steady incremental accumulation of minor changes resulting from simple gene replacement through natural selection. Instead, it proposed that speciation was a relatively rapid process, occurring in what on a geological scale amounts to a geological instant.

> I make the operational suggestion that it be defined as 1 percent or less of later existence in stasis. This permits up to 100,000 years for the origin of a species with a life span of 10 million years, *though I believe that most events of speciation occur much more rapidly*. (Gould, 1982a: 84, emphasis added)

Moreover, stasis,—the "relative lack of significant phyletic change within species once they appear,"—marks the rest of their existence as species (Eldredge, 1989: 70). In other words, species are not in some constant gradual state of phyletic becoming, without ever actually existing as discreet entities for some discreet interval. They come into being relatively rapidly, exist for some protracted length of time in a relatively unchanged state *with respect to their defining characteristics*, and then die out.

True, for any given species, its beginning is something of a 'fuzzy event,' for not being truly instantaneous. However, that fuzzy beginning eventually ends, and the species then exists. More to the point, the 'fuzziness' of their beginnings does not inhibit us from comfortably accepting organisms or celestial bodies as individual entities (e.g., see Hull, 1976, 1980; Sober, 1984). When does an organism, such as an individual dog, cat, or human being 'begin?' Is it when the egg is fertilized, when it leaves the womb, or at what point in between? At what point in the condensation of a gas cloud does it become a planet or a star? Is it when the first particle of interstellar dust is attracted to the second, when the last one falls into the whole, or at some point in between, as when a body reaches some conception of minimum planetary size or when a star first reaches sufficient internal pressure to ignite? Yet for all the fuzziness of such beginnings, we have no conceptual trouble considering both individual organisms and celestial bodies to be discreet entities because they *have a beginning* and remain relatively unchanged with respect to their defining characteristics as entities subsequent to their beginnings.

However, stasis does not imply the complete absence of any kind of change. Rather, it is defined as stability with respect to just the basic morphotype. Within that morphotype, microevolution that tracks environmental changes continues unabated.

> The stasis found in morphological characters in [certain] old genera . . . is not at all due to the retention of an entirely unchanged genotype. . . . [C]ountless changes in quasi-neutral enzyme genes have been discovered [and] numerous other non-morphological changes have also taken place in these genera . . . *What has remained stable, however, is the morphotype, the basic Bauplan.* (Mayr, 1989: 146, emphasis added)

The key to considering species (or individual organisms and planets, for that matter) as individual entities is not by focusing on their actual beginnings, with all its attendant fuzziness. Rather, the key is in recognizing the fact that *the beginning ends at some point*; and that beginning is then followed by some significant period of existence in a relatively unchanged state with respect to the species' basic structural characteristics. Those basic structural characteristics, its Bauplan (see also Eldredge, 1989: 43–46), are not the essence of the species. They are simply the basic structural characteristics shared by its constituent members, by virtue of common inheritance. However, members are not conspecific by virtue of genotypic or phenotypic similarity, as in similarity of Bauplan; they are conspecifics by virtue of their historical (reproductively derived) relationships to each other and from which the similarity springs (see Ghiselin, 1974; Hull, 1976, 1978; Sober, 1984). "Species . . . are chunks of the genealogical nexus, not natural kinds" (Sober, 1984: 165; see also Eldredge, 1985a: 161), nor are they points on a continuum.

As noted by Hoffman (1989), Ruse (1989), as well as Gould (1989a), punctuational evolution was initially presented (Eldredge, 1971; Eldredge & Gould, 1972) as simply an alternative to phyletic gradualism. The focus was on the pace of speciation, not on the processes that produced it. Thus, biological evolution as a process of "punctuated equilibria" could readily be reconciled with other elements of the neo-Darwinian synthesis by simply postulating periodic bursts of more rapid incremental change. It is even possible to legitimately claim (e.g., Hoffman, 1989: 187; see also Mayr, 1989) that the essential points were already implicit in elements of the Modern Synthesis (e.g., Simpson, 1944).

However, the theoretical ramifications of their theories also forced the reemergence of a "taxic perspective—the acceptance of the existence of species as discreet entities, with births, histories, and deaths. . ." (Eldredge, 2015: 204), particularly as it relates to species selection/sorting, hierarchy, adaptation, and mutation.

This ultimately led Gould (1980, 1982a, 1982b, 1989a, 2002a), Eldredge (1985a, 1985b, 1989, 2015), Stanley (1979, 1981, 1989), and others (e.g., Vrba, 1983, 1984, 1989) to increasingly diverge from the Modern Synthesis, particularly with regard to its then reductionist direction and by further extension with regard to the role of natural selection in speciation. That is, speciation came to be viewed by them as a rapid and *discontinuous* process, that the evolution of species results not so much from a living system in operation as from a systemic reorganization (e.g., Gould & Eldredge, 1977: 144). In contrast to the neo-Darwinian synthesis and the genetic selectionism that evolved to dominate it, the evolved punctuational perspective accords adaptation a limited role in evolutionary change at any level beyond the microevolutionary. Instead, punctuationists view natural selection as a force that promotes evolutionary stasis within species while acting on entire species at a higher, macroevolutionary level (e.g., Eldredge, 1989). Thus, biological evolution is proposed to be hierarchical, with speciation and macroevolution involving processes distinct from those involved in microevolution.

Macroevolution or group/species selection refers to the process said to underlie evolutionary trends. Its operation is implicit in punctuational theory.

> . . . [I]f species originate in geological instants and then do not alter in major ways, then evolutionary trends cannot represent a simple extrapolation of allelic substitution within a population. Trends must be the product of differential success among species. (Gould, 1980: 125)

> What I mean by success of a species is the displaying of one or both of two traits: first survival for a long time, and, second, rapid production of descendant species that share the species' basic attributes. (Stanley, 1981: 182)

Thus, if species are themselves inputs "and [evolutionary] trends the result of their differential origin and survival," there must then exist "a higher form of selection, acting directly upon species through differential rates of extinction," in a manner analogous to natural selection operating on individuals within populations (Gould, 1980: 126; see also Eldredge, 1989: 139ff). Like individuals, it is argued that species do not change significantly in response to long-term environmental change; they either survive with only relatively modest change or die out (see also Eldredge, 2002).

> Accordingly, environmental factors do not so much direct change as limit it.
>
> While environmental agents seldom drive evolution very far within large well-established species. . ., they must often dictate the extinction of species. . . . They must also frequently determine whether or not small populations blossom into new species and where and when speciation occurs. (Stanley, 1981: 187)

Therefore, as first argued by Stanley (1975: 648), macroevolution is not simply microevolution writ large. Rather, proponents consider it to be governed by both different stochastic and selective processes than those that drive microevolution.

Biologically, novelty can only arise from the fixing of a mutation by selection. Major evolutionary changes by (gradualist) definition require the individual fixing of a succession of mutations. Gradualists explain the requisite intermediate stages of major transitions by recourse to what used to be called preadaptation and is now called exaptation. However, if evolution is hierarchical and governed by distinct processual modes at the macroevolutionary level, proponents raise the possibility that speciation is governed by yet a third mode of change.

> I do not doubt the supreme importance of preadaptation, but the other alternative ... now deserves a rehearing...: perhaps, in many cases, the intermediates never existed. (Gould, 1980: 127)

Gould did not argue for the saltational origin of entire new designs, complete in all their complexity, as per the discredited genetic theories of evolution replaced by the neo-Darwinian synthesis. Instead, he proposed a discontinuous origin for the "essential features of key adaptations," which "might set up new regimes of development and selection that would quickly lead to other, coordinated modifications" (1980: 127). In other words, the morphological *equivalent* of limited macromutation can easily arise as a consequence of simple mutations in regulatory genes (e.g., Gould, 1980: 127, 1983: 196; Stanley, 1981: 135).

> I do feel that certain forms of macromutational theory are legitimate.... Legitimate forms include the saltatory origin *of key features* (around which subsequent adaptations may be molded) and marked phenotypic shifts caused by small genetic changes that affect rates of development in early ontogeny *with cascading effects thereafter*. (Gould, 1982a: 88–89, emphasis added)

Given this proposed potential for simply produced dramatic change, an important possibility follows. That is, the possibility then exists for random genetic processes coupled with other possible factors (e.g., see Gould, 1983: 336ff.) to produce reproductive isolation and hence speciation independently of natural selection. In such cases,

> [R]eproductive isolation comes first and cannot be considered an adaptation at all. It is a stochastic event that establishes a species by the technical definition of reproductive isolation.... We can, in fact, reverse the conventional view and argue that speciation, by forming new entities stochastically, provides raw material for selection. (Gould, 1980: 124)

Thus, speciation, the process that produces the raw material for macroevolution, is proposed to be governed by stochastic processes different than those producing the raw material for microevolution. Moreover, such reproductive isolation has been documented to occur in as little as 8 to 13 generations (Hendry et al., 2000), a 'geological instant' by any measure.

However, if at the level of speciation, selection arguably operates on existing *constellations of key features*, then other traits are arguably of secondary importance. Beyond the retention of phyletically inherited genetic baggage, this opens the door to the possibility that many traits may simply be new genetic baggage produced during speciation, just "the necessary consequences of materials and designs" inherent in the basic organic structure or "Bauplan" of successful organisms (Gould & Lewontin, 1979: 595). Thus, adaptation via natural selection is accorded a limited role even in the "fuzzy beginnings" of new species.

The neo-Darwinian perspective readily acknowledged that traits can exist by virtue of pleiotropy, allometry, selective trade-offs, and other factors only indirectly related to adaptation. However, Gould and Lewontin (1979; see also Gould 1980,

1982a, 1982b, 1989a) argued that adaptively irrelevant (and perhaps even maladaptive) traits are potentially much more numerous than generally acknowledged and often go unrecognized as such in the search for an adaptationist explanation. More importantly, they suggest that the Bauplan and consequent traits, to the degree that they constrain change, are as important in determining the direction of evolution as the selective forces that mediate change.

> Selection may supply an immediate direction, but if highly constraining channels are built of nonadaptations, and if evolutionary versatility resides primarily in the nature and extent of non-adaptive pools, then "internal" factors or organic design are an equal partner with selection. (Gould, 1982b: 384)

These constraints of organic structure (Bauplan) took Gould full circle by explaining stasis as the product of constraints instead of stabilizing selection.

> I continue to have one cardinal difficulty with the attribution of stasis to stabilizing selection—the scales are all wrong.... I would rather suppose that the profoundness and temporal depth of stasis are trying to tell us that change is *actively prevented*, rather than always potential but merely suppressed because no adaptive advantage would accrue. (Gould, 1989a: 124, emphasis in original)

The theoretical ramifications of punctuational theory remain controversial to one degree or another (e.g., see Shostak, 2003), particularly in the eyes of genetic selectionists. However, Mayr (1989: 145), one of the architects of the Modern Synthesis and something of a critic of punctuational theory, noted that the facts are that whether or not "pronounced stasis [is] the usual fate of most species," as claimed by Gould (1982a: 86), it has clearly been the fate of many species. Moreover, Mayr (1989: 145–147) also dismissed conventional explanations for stasis as inadequate. Thus, for much the same reason as the proponents of punctuated equilibria, he concurred with their anti-reductionist view of macroevolution and recognized the necessity for some kind of 'genetic revolution' if speciation is to occur (see Mayr, 1954, 1982, 1989).

> ... [Reductionist genetic models can be offered that] explain everything in terms of the simplest single-gene assumptions. These models cannot be refuted, but they leave far more natural phenomena unexplained than the theory of a genetic restructuring of founder populations. (Mayr, 1989: 149)

Of equal importance, Mayr asserted the compatibility of the hierarchical *essence* of punctuational theory and modern neo-Darwinism.

> Holists [in which category Mayr places himself] claim therefore that much of macroevolution cannot be explained by atomistic gene replacement or by selection pressures on single genes, but only by a more drastic reorganization.... It has been claimed that this holistic view of the genotype was not 'within the spirit of modern Darwinism'. This is erroneous. The atomistic viewpoint was defended only by mathematical geneticists.... (Mayr, 1989: 147)

Macroevolution and Group Selection

Even if the exact mechanisms involved in the rapid speciation events integral to the theory of punctuated equilibria are still not universally agreed upon, the relatively long duration of stasis and the relative speed of such speciation events eventually tended to largely be accepted (e.g., Mayr, 1989; Hoffman, 1989). Thus, sidestepping the issue of the exact processes that produce new biological species, species can be seen as fully distinct entities on the basis of pace alone. The fact that species are spatially bounded synchronic entities was never in serious doubt. The speed of speciation events allows them to also be considered historical entities, diachronically bounded with distinct beginnings (however fuzzy) and ends—in some potential hierarchy of evolutionary entities only beginning with the gene. However, to be evolutionary entities, species must be subject to selection, and such selection must be some form of group selection.

The concept of group selection dates back to Darwin (1871) himself. As originally conceived, the term was typically applied to those attributes of individuals that were thought to be selected for because they benefitted the group (e.g., deme, species) to which the individual belonged, at the expense of the individual organism. The evolution of altruistic behaviors on the part of individuals was the classic case of a characteristic that logically suggested the possible operation of such (for-the-good-of-the) group selection on individual organisms. That is, altruism benefits the group that the altruist belongs to but costs the altruist, thus implying the operation of something other than organismic selection—namely, (for-the-good-of-the) group selection—for its perpetuation.

However, over time, four powerful arguments against such group selection were advanced to demonstrate the impotence of such (for-the-good-of-the) group selection in the face of organismic selection (see Fisher, 1958; Williams, 1966; Hamilton, 1971, 1987). For example, organismic selection allows cheaters to outcompete altruists within the group because they reap the benefits of altruism by others without incurring the cost of being altruistic themselves (see Gould, 2002a: 646–647 for a concise summary of all these arguments against group selection). In a very real sense, sociobiology (Wilson, 1975) and the concepts of kin selection and inclusive fitness, as well as the concept of reciprocal altruism (Axelrod, 1984), were eventually advanced specifically to offer viable alternative explanations for the evolution of altruistic behaviors at the organismic level in the face of that impotence (e.g., see Laland & Brown, 2002).

However, at about the same time as sociobiology and reciprocal altruism were being put forward as alternative explanations to group selection for altruism, the concept of group selection itself was being resurrected in two slightly different ways (e.g., see Damuth & Heisler, 1988; Brandon, 1988, 1990). With respect to the 'classic' debate, centered on the attributes of individuals that conceivably evolved for the good of the group at the expense of the individual, as with altruism, counterarguments were advanced (see Wade, 1978; Sober, 1984; Wilson & Sober, 1994; Sober & Wilson, 1998; Joyce, 2006). To begin with, modeling demonstrated that under several sets of plausible conditions, the declining intra-demic relative frequency of individuals with a given good-for-the-group characteristic (e.g., altruism) produced by organismic selection can be overcome by the relative *inter*-demic success of demes having such members. Thus, the "overall frequency [of individuals with that attribute] may rise within the species even while their frequency within each surviving deme declines" (Gould, 2002a: 648). Add to that the fact that the arguments against (for-the-good-of-the) group selection were implicitly based on natural selection being the only significant force operating on individuals, with sexual selection not entering into the equation. They thereby fail to consider any possible reproductive advantages such individuals may gain due to their good-for-the-group attributes that potentially outweigh the somatic disadvantage imposed by that characteristic (e.g., see Zahavi & Zahavi, 1997). Moreover, some well-documented patterns in nature (e.g., significantly female-biased sex ratios in some species) seem very difficult to explain without recourse to some significant level of such inter-demic selection (see Wilson & Sober, 1994; see also Gould, 2002a: 648–649 for a concise summary).

More importantly for present purposes, the theory of punctuated equilibria fostered a growing interest in macroevolutionary theory, hierarchies of entities, and species (as opposed to inter-*demic*) selection/sorting[1] (e.g., Stanley, 1975, 1981; Vrba, 1980; Eldredge, 1985a, 1989, 1999; Gould, 2002a). The key point regarding the demic-species distinction is that the classic arguments against group selection "had legitimate force" in the matter of inter-demic selection, "but could be overcome under [certain] conditions" (see above). However, in the matter of species selection, they either do not apply at all or become irrelevant (see Gould, 2002a: 649 for the full argument).

Of even greater importance for present purposes is the point that group selection is explicitly viewed by the punctuationalists/macroevolutionists as operating on the properties of the group as a whole—, e.g., differential susceptibility to speciation (see Eldredge, 1999; Gould, 1982a, 2002a; see also Vrba, 1980 for a good example). Thus, in a very real sense, the macroevolutionists are proposing a very different concept of group selection than was and still is the subject of the ongoing 'classic' debate concerning altruism. That is, the original concept of group selection focused on the characteristics of individual organisms. Prior to Fisher's (1958) critique of group selection, the focus of the 'classic' debate was on the aggregate bottom-up, 'good-for-the-group' effect of individuals' attributes on the success of the group. In its resurrected form (i.e., Wade, 1978; Sober, 1984; Wilson & Sober, 1994; Sober & Wilson, 1998), the focus has shifted to include the concurrent, offsetting top-down effect on individuals of membership in groups enhanced by the presence of such individuals; but the focus remains on the attributes of individuals, as members of

[1] "Selection is a claim about causality—differential births or deaths based upon characters of the objects being sorted. "Sorting" is the descriptive claim that such a differential exists, leading to the accumulation of some trait or other within a population or lineage" without reference to the cause of that differential (Gould, 1989a:122; Vrba, 1980, 1984, 1989; see also Eldredge, 1989:140ff for a discussion of the distinction).

groups. Some macroevolutionists refer to this form of species sorting as produced by the bottom-up "effect hypothesis" (Vrba, 1980; see also Eldredge, 1989: 144ff) and not true species 'selection.' In contrast, the macroevolutionists have focused on groups, particularly species, as integrated individuals visible to the operation of selection as higher-order entities and their relative success (e.g., geological longevity, speciational 'success') in what amounts to a variant of typical Darwinian competition, but between groups, at the group level. In other words, the focus here is much more strongly fixed on top-down effects that have no bottom-up basis in the specific attributes of the group's constituent individual organisms.

As noted by Gould, these "two approaches often yield concordant results for the obvious reason that differential proliferation of higher-level units . . . often defines the group effect that influences the fitness of lower level individuals." But this need not always be the case, "*leading to situations where we could identify group selection by one criterion, but deny [it] by the other*" (2002a: 653, emphasis added; see also Dugatkin & Reeve, 1994).

Lastly, it is necessary to clarify just what between-group selection is said to operate on in this hierarchical, macroevolutionary, top-down approach to group selection and on what basis it is said to operate (see Grantham, 1995; Eldredge, 1989: 140ff; Gould, 2002a: 659ff). One view of the basis for such top-down selection (Vrba, 1983, 1984, 1989; Vrba & Eldredge, 1984; Vrba & Gould, 1986) focuses on "emergent character." That is, it looks to the emergent systemic properties of groups as the locus for the operation of selective forces on groups as higher-order individuals. Just as individual organisms are systemic individuals with distinct emergent properties, resulting from non-additive genetic interactions that do not exist at the next lower level, groups are likewise systemic individuals with potential emergent properties (resulting from the non-additive interactions of individual organisms) that do not exist at the level of their individual constituent organisms. And selection operates on individuals of whatever order based on their variable attributes as individuals of that order. For example, a social species' mode(s) of group organization may be an interactive product of individual drives encoded in the constituent individuals' genes, but the organization itself is an interactive feature of the group, not of any individual. And to the degree that there is variability between groups in the specifics of their organization, selection can be assumed to operate on those groups on the basis of their variable organization.

The alternate view (Lloyd, 1988; see also Lewontin, 1970; Arnold & Fristrup, 1982; Damuth, 1985; Damuth & Heisler, 1988) focuses on the more inclusive concept of "emergent fitness," thereby including additive, sum-of-parts, as well as emergent group-level characteristics. For example, the aggregate variability within a group for a given trait is an additive sum of parts. Since the appearance of a variant must precede selection for that variant and the appearance of variants is independent of any given variant's usefulness, existing variability determines the number of avenues open for evolution. Thus, in evolutionary terms, for a given group, higher variability is superior to lower variability because it constitutes more open evolutionary 'avenues' for selection to potentially produce an adaptation to changing conditions. This is so, even though it is obviously not superior for those individuals who have anything less than the optimal variety of a given trait at any point in time.

As noted above, these two views are not in opposition. Rather, "Vrba's exclusive [emergent character] domain [constitutes] a subset of 'best cases' in Lloyd's [emergent fitness] formulations" (Gould, 2002a: 659). But that still leaves the question of what kind of groups between-group selection is said to operate on, and this is a complex issue that takes us full circle back to the question of what a species is said to be.

Selection and Hierarchy

As noted previously, the only *meaningful* view of species is reproductively based, since species are not always monolithically integrated into ecological systems and hence cannot consistently be defined economically. Yet species display distinct phenotypic adaptations, however general, and not all of these relate to reproduction. Some are economic. Thus, the murky but very real ecological aspects of species continued to lurk in the background as an unresolved issue in the matter of group selection. That is, if species are considered (higher-order) reproductively defined individuals (i.e., replicators) on which selection operates, as per the macroevolutionists' view, then they are entities in some potential hierarchy of evolutionary entities only beginning with the gene (e.g., see Ghiselin, 1974; Paterson, 1981; Eldredge, 1985a, 1989, 1999). But natural selection operates on interactors, not replicators. Thus, the ecological aspect of species needed to be clarified and reconciled somehow with the species concept as reproductively defined, if species were to be considered legitimate evolutionary entities.

Ecological interactors have long been recognized as being hierarchically organized. Thus, initial attempts to reconcile the reproductive definition of species and their ecological aspects were in the form of hierarchies that tried to integrate both the reproductive (evolutionary) and economic (ecological) aspects of species within a single unified hierarchy (e.g., Valentine, 1973; MacMahorn et al., 1978). However, in each such case, the results were less than fully satisfactory, if for no other reason than the stubborn inability to characterize species in ecological terms (see Eldredge, 1985a: 165ff for specifics). This prompted Eldredge (1985a, 1989, 1999; see also Eldredge & Salthe, 1984; Vrba & Eldredge, 1984) to model evolutionary dynamics in terms of two separate, but dynamically intersecting and interacting, biological hierarchies that neatly eliminated much of the conceptual conflicts regarding evolutionary and ecological units (Fig. 2.1).

The first is a genealogical hierarchy whose components are involved in information transfer and have in common the capacity for "more-making" (Eldredge, 1985a: 144). Genes make more genes, and chromosomes make more of the same, as do individual organisms, demes, and species, up to monophyletic taxa; and in the process of differentially so "more-making," they evolve. Units within this genealogical hierarchy are characterized by reproductive adaptations, and their evolution

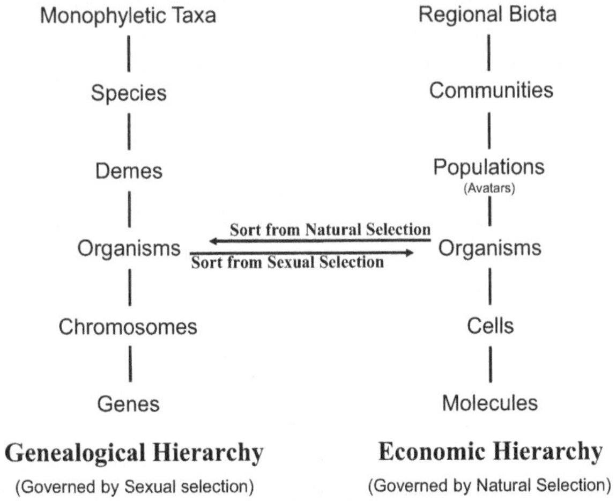

Fig. 2.1 Eldredge's system of dual biological hierarchies (after Eldredge, 1985a, 1989, 1999)

is propelled by reproductive interactions,—sexual *or otherwise*. That is, we can define the mode of selection broadly as favoring "relative reproductive success based solely on differential reproductive attributes," without restricting it to just the product of competitions inherent in the sexual reproduction of individual organisms (Eldredge, 1989: 203). If so, then all the entities in this hierarchy, up to and including species, which make more species by speciating, evolve by means of what we can only very broadly continue to call 'sexual selection' because it includes various asexual reproductive analogs (e.g., see Stanley, 1975). More appropriately, it should probably be called reproductive selection, but that issue is not worth pursuing by an archaeologist like myself least of all.

The second is an ecological/economic hierarchy whose components are defined by level of interaction involving matter-energy transfers—economic interactions—between interactors. Molecules interact with other molecules as well as the environment. So do cells, individual organisms, populations (avatars), and communities consisting of local communities/avatars of diverse species, up to regional biotas; and it is these interactions that drive natural selection, thus providing the raw material for sexual selection to act on. "It is the function of ... the genealogical hierarchy to supply... 'players'" to the economic hierarchy (Eldredge, 2002: 318), and it is the function of the economic hierarchy to supply "players" to the genealogical hierarchy (see also Eldredge, 1985a, 1989, 1999).

Distinct dynamics select from among the entities within each hierarchy because each hierarchy is characterized by a distinct mode of selection. In the economic hierarchy, it is natural selection, and it is sexual selection in the genealogical hierarchy. Sexual selection in the genealogical hierarchy acts on what the ecological hierarchy allows to pass in the form of reproductive attributes, and natural selection in the economic hierarchy acts on what the genealogical hierarchy allows to pass in the form of economic attributes.[2] The sort from one hierarchy passes through the nexus of the individual organism, which level of entity straddles both hierarchies, to provide the raw material for the selective forces operating within the other hierarchy. In other words, natural selection determines which individuals are available for potential reproduction on the basis of their economic attributes (but who also incidentally have reproductive attributes), and sexual selection determines which individuals reproduce on the basis of their reproductive attributes (but who also incidentally have economic attributes) for natural selection to operate on.

Individual organisms function as entities within both hierarchies and constitute the nexus of the two hierarchies. They thus act as both replicators and interactors and are thus subject to both natural and sexual selection. This existence of individual organisms as entities within both hierarchies is arguably the primary source of the conflicted views of species. That is, species are not truly economically adapted; but the impression that they are so adapted is falsely projected by the fact that their component organisms are—but not uniformly so—as individuals within the economic hierarchy. Species do not occupy niches in any meaningful way; local populations (not to be confused with demes) do. Species, as reproductive units (which are also reservoirs of economic as well as reproductive information), fall within the genealogical hierarchy. Local populations, which are the spatially variable embodiment of that economic information, fall within the economic hierarchy. Species, however, evolve and populations do not.

Lastly, a point needs to be made clear. Eldredge (2002: 329) likens the internal dynamics of these dual hierarchies to the motion of liquids in a "sloshing bucket" (Fig. 2.2), with the sorts from each hierarchy "moving" to the other at the bottom of the bucket—the individual—and sometimes in the face of sufficiently disruptive events sloshing up the sides of the bucket to affect higher-order entities. While he used the metaphor specifically to illustrate the scalar effects of environmental change on the entities in his hierarchies, the metaphor also illustrates an important point about group selection.[3] That point is that between-group selection is not something that theoretically raises the bucket's floor—and hence the level of the 'sloshing' sorts—to some higher level of entities above the individual. Firstly, this is because the competitions are only between similar-order groups in a given hierarchy. Economic entities compete only with other economic entities and not with reproductive entities. The same kind of exclusion applies to genealogical entities. More importantly, this is because while intergroup selection involves competition between groups, it is materially experienced firsthand only by the groups' constituent individuals through the top-down effects of membership. That is, if all the individuals in a group cease to exist, that group then ceases to exist, and if all the groups that

[2] An organism can be supremely adapted to its economic environment, but if being so makes it supremely unattractive to potential mates, or hampers reproduction in any other way, its fitness still suffers greatly. The same applies to any negative economic effects of reproductive adaptations.

[3] Here being used as a general term that includes both group and species selection (see Eldredge, 1989:140ff; Gould, 2002a:659ff for discussions of the difference between group and species selection).

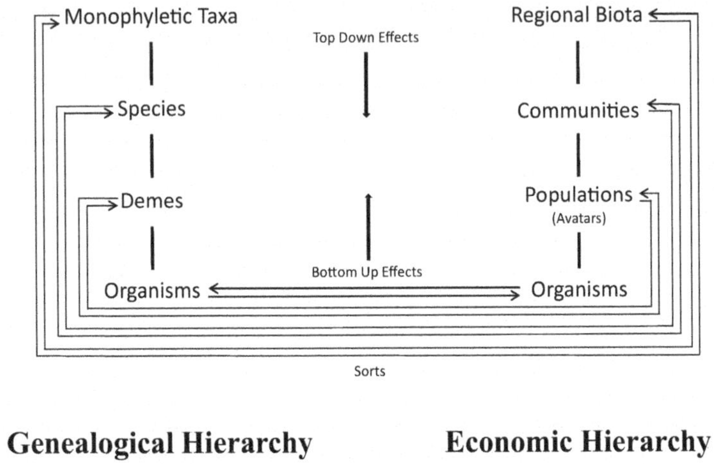

Fig. 2.2 Eldredge's 'sloshing bucket' notion of evolution, modified to show group selection (after Eldredge, 2002)

constitute a higher-order group cease to exist, *because all their respective constituent memberships have ceased to exist*, that higher-order group then ceases to exist. But the selective effect is a product of competition between groups on the basis of both their group level attributes and those attributes' effects on the group's constituent members, as well as the group's sum of parts' effects.

Not all biological evolutionists who see species as bounded entities accept Eldredge's approach based on a system of *dual* hierarchies. For example, Gould (2002a: 642), who also favored a genealogical view of species and the existence of multilevel selection, objects to a system of dual hierarchies on the basis of what he sees as the theoretical necessity for "units of selection [to] be evolutionary individuals." However, that necessity seems to stem as much from his desire for theoretical elegance as from theoretical logic.

> I find the idea of dual hierarchies both interesting and challenging, but ultimately flawed and counterproductive in the introduction of unnecessary complexity. (Gould, 2002a: 642)

Eldredge's dual hierarchies may be more complex than some hypothetically workable unified hierarchy, but Gould offered no specific alternative; and so the claim of "unnecessary" remains to be demonstrated.

On the other hand, Damuth (1985) argues that ecological entities, particularly his "avatars" (i.e., local populations), are capable of 'more-making' and that a separate reproductive hierarchy is therefore unnecessary. However, while Damuth's avatars can occasionally divide, it is an inconsistent form of reproduction (see Eldredge, 1989: 186), and expanding such an approach to its logical conclusion would take the 'species issue' back to square one. In the interim, a system of dual hierarchies offers the most comprehensive solution to reconciling the ecological and reproductive aspects of biological species. As will be explored later, it does so equally well,—if not more so (if further expanded),—when applied to cultural entities, but it does so at the expense of still greater complexity. This is because, as will also be explored later, when dealing with culture, there are more than just the two biological hierarchies that must be considered. There are also cultural hierarchies that must be considered, whose interplay with each other and the biological hierarchies has driven the coevolution of humans and culture.

While the exact mechanisms that might produce punctuational speciation remain controversial, the prevalence of stasis and the relatively prolonged duration of such stasis are considerably less so because that aspect of punctuational theory deals solely with pace and could thus be much more readily reconciled with the neo-Darwinian synthesis. However, the biological mechanics producing punctuational speciation are irrelevant to cultural evolution because the mechanics of cultural evolution operate on entirely different entities than those that are subject to biological evolution—metaorganic ones (or if one prefers, superorganic) versus organic ones. Those entities are structured and transmitted differently than those that are involved in biological evolution—being based on learning versus genetic inheritance—and are subject to different evolutionary processes, such as Lamarckian mechanisms operating alongside Darwinian ones. The point is that the controversy concerning the actual mechanics of punctuational events in the biological domain is completely tangential to any discussion of the mechanics underlying the possibility of any such events in the cultural domain because they are inevitably going to be different by simple virtue of the fact that biological and cultural entities are structurally different. Such cultural evolutionary mechanics thus need to be evaluated on the logic of their own internal linkages, on their viability to produce the claimed effect, and most importantly on their degree of consilience with the larger accepted body of culture theory, evolutionary or otherwise.

Epigenetics, Organisms, Niche Construction, and Agency

Lastly, there are a number of other biological tracks whose implications began to converge with the theoretical fallout from the Paleontology-driven view of evolution as a process of punctuated equilibria and particularly its consequent focus on higher-order entities than genes. What they had in common was the view that the genetic reductionism that had come to dominate evolutionary theory by the 1970s was unworkable (e.g., see Dayan et al., 2019 for one relatively recent such discussion). What they also had in common besides the shift away from genetic reductionism was a return to Darwin's organism-centered view of evolution—a return to the importance of the individual.

In general, by the end of the twentieth century, the developing field of molecular genetics was casting increasing doubt on the viability of genetic reductionism. That is, it was becoming increasingly clear that genes simply do not work the way they are required to work for the single gene replacement model inherent in the reductionist view of evolutionary dynamics that held sway. Selective forces can only operate on the phenotypic manifestations of genetic traits, and many if not most of an organism's phenotypic features that are most visible to selective forces are the product of genetic interactions as opposed to individual genes, with genes often involved in multiple diverse such interactions. Thus, the required ruthless operation of selection on individual genes is seriously hindered by individual genes' obscurity to selective forces due to their being part of the larger interactive systems that are the phenotypic features they contribute to and the fact that they can contribute to multiple diverse phenotypic traits.

More importantly still, Gould's (1977, 1982a) call to refocus attention on the relationship between ontogeny (individual development) and phylogeny (evolution) was part of a trend toward increasing attention being paid to what Waddington (1940) had termed "epigenetics" and the eventual development of what has come to be called Evolutionary Developmental Biology (see Hall, 1992, 2012). Epigenetics was perhaps best defined by Hall (1992: 89) as "the sum of the genetic and non-genetic factors acting upon cells to control selectively the gene expression that produces increasing phenotypic complexity during development." The first key point here is that by the end of the twentieth century, it was becoming quite apparent that some such nongenetic factors were heritable. As summed up by Maynard-Smith (1989: 11), "there is a second inheritance system—an epigenetic inheritance system—in addition to the system based on DNA," which affects the expression of DNA. The second key point here is that development is something that organisms undergo and genes do not.

On another front, Dawkin's (1976) agential metaphor of the "selfish gene" notwithstanding, genes have no self to be selfish; thus, genes cannot be agents, and agency cannot be an aspect of evolutionary processes in the view of genetic reductionists. While such "agential thinking" (Godfrey-Smith, 2009; Okasha, 2018), in this case concerning "selfish" genes, can usefully summarize adaptive insights functionally, it is not the same as actually ascribing agency.

In the reductionist view, organisms are merely avatars of and vehicles for the propagation of "selfish" genes, and this had a broadly pernicious conceptual effect. That is, from this perspective, evolutionary entities (i.e., genes) are merely seen as being passively acted upon by selective forces. In other words, evolution is simply something that happens to species as a result of selective forces producing changes to their constituent organisms' genetic properties and to which forces they remain passive. From a genetic perspective, such a view is perhaps justifiable due to the stochastic nature of the processes producing genetic novelties—i.e., mutation. However, true agency is intentional, purposive action, as opposed to simply behavior. It is the doing of something that has a particular etiological basis (Okasha, 2018: 12–13). Genes are incapable of agency, but organisms are quite capable of it, to which I would add that the more sentient the organism, the more potentially and complexly so capable. For organisms, behavioral novelties are not necessarily restricted to ones produced specifically by underlying heritable biochemical changes, with preexisting phenotypic plasticity being an obvious source of such non-biochemically produced behavioral novelties.

The point here is that concurrently with the growing recognition of the importance of ontogeny in evolutionary processes, there was a growing recognition (e.g.,

Lewontin, 1982, 1983; Odling-Smee, 1988) of the fact that organisms are not just passively subject to the operation of selective forces. Rather, they act and by doing so often enough modify those very selective forces. This crystalized within Evolutionary Ecology in the form of Niche Construction Theory (e.g., Odling-Smee, 1988; Lewontin, 2001; Odling-Smee et al., 2003; Laland et al. 2019). As noted by Lewontin (2001: 64), "the environments of organisms are made by organisms themselves as a consequence of their own life activities." That is, by engaging in life activities, organisms modify their environments as a consequence of those activities, which environmental modifications in turn modify the selective regimes affecting both the species responsible for the modifications and other species as well. In other words, environments coevolve with organisms (see Odling-Smee et al., 2003: 2). Moreover, those life activities can be agential.

The case for organismic agency's role in evolution was best made by Walsh (2015), building on the insights of both Evolutionary Developmental Biology and Evolutionary Ecology. For Walsh, adaptation is not a "process in which the ... environment moulds passive form. Rather, it is a process by which organisms respond to, and in the process create, their own system of affordances" (2015: 164), an affordance being "a joint property of a purposive system and the conditions with which it interacts" (215: 163). In a manner of speaking, an affordance is a perceivable action possibility, and such possibilities obviously vary with both the properties of the agent and the properties of the environmental conditions surrounding the agent.

The basic problem such a view creates is that organismic agency cannot be accommodated by the reductionism of the Modern Synthesis and its structural reliance on organismic passivity in the face of selective forces.

> The most acute defects of the Modern Synthesis issue from its marginalisation of organisms and its excessive reliance on genes.... Organisms are different from genes. Genes, on the Modern Synthesis view, are mechano-computational devices.... *Organisms are purposive*, self-synthesizing, self-regulating entities, open systems, constantly exchanging matter and energy with their environments. (Walsh, 2015: 163, emphasis added)

Walsh argues that the root of the problems with the Modern Synthesis is that evolutionary processes (e.g., selection, mutation, etc.) became "fractionated"—i.e., treated as being discretely independent of each other—in the Modern Synthesis and that in reality they do not operate quite as independently of each other as they are held to do in the Modern Synthesis. Rather, "evolution is a consequence of agency—organisms' purposive engagement with their affordances" (2015: 227). In place of the bottom-up reductionism of the Modern Synthesis, Walsh proposes an alternative that he calls Situated Darwinism, which places at least equal importance on top-down effects.

> An overreliance on the bottom-up strategy of mechanism obscures the explanatory reciprocity between parts and wholes. ...The parts of organisms get their particular causal powers from their contexts—the activities of the organisms as a whole.... (Walsh, 2015: 219)

> The crucial difference between evolution construed this way and the Modern Synthesis approach is that, according to the latter, evolution is fundamentally a molecular

> phenomenon. According to Situated Darwinism, evolution is fundamentally an ecological phenomenon. (Walsh, 2015: 163)

Thus, Walsh's view of evolution, at least to the degree that it places the focus on higher-order entities, such as organisms, and more importantly on interactions, arguably dovetails with Eldredge's view of hierarchical evolutionary dynamics, with their selective interactions and the sloshing bucket having individual organisms as the nexus of those interactions.

Logically, an organism must be doing something in order for selective forces to promote physical and/or behavioral traits that facilitate the doing of that something over the doing of something else. More importantly, the doing of that something may occur in the absence of genetic instructions mechanistically programming the doing of it, and the doing of it may be truly agential. Whether an individual example of the doing of a specific something can best be explained by just abstract agential thinking or by the ascription of true agency and specific goal orientation is an unanswerable question as concerns most species. An animal's mind is (for now at least) truly a black box. We may never know whether, for example, a male lion engaging in infanticide is doing so specifically with the intent to induce estrus in the killed cubs' mother or just because, for example, he is just lethally hostile to any other lion of whatever age that does not exhibit the correct visual or olfactory traits, with the induction of estrus being simply a selectively valuable by-product of killing youngsters that are not his own. Thus, the prudent explanatory course in most cases is just agential thinking related to adaptive function. But the fact remains that whatever the specific motivation for the action may be and whether or not the trigger for that action is genetically based, the (in this case) male lion does have a motivation and goal orientation and is thus actually acting agentially, even if we cannot know definitively whether the selected-for functional benefit was what specifically motivates him to bring it about and must therefore prudently restrict ourselves to only agential thinking.

The point here is that the answer to whether we can ascribe agency or must restrict ourselves to just agential thinking lies in the degree to which we can look into the black box. Humans are arguably the most complexly agential organisms on the face of the planet. This is attributable in large part to our large brains and its ability to remember and process information and also to our vast speech- and observationally produced accumulated store of transmitted information. What that means is that for humans, not all behavioral motivations are of necessity equally obscure and so, under the right circumstances, we can sometimes look into the black box and ascribe purposeful agential motives to a behavior. More importantly, we can sometimes, under the right circumstances, arguably claim to look into the black box even for long-dead group-level human entities, and this is an issue we will return to later when we come to the issue of cultural evolution.

Chapter 3
Borrowing from Biology

The most acute defects of the Modern Synthesis issue from its marginalisation of organisms and its exclusive reliance upon genes as the canonical unit of evolution.
Denis M. Walsh

Evolution and Culture

It is a generally accepted fact that the human capacity for culture has evolved from some earlier, more primitive state of lower capacity akin to that which characterizes other hominids. This is evident in both the increasing size of the human brain over the course of hominin biological evolution and the increasing elaboration of cultural behavior over that same span (cf. Durham, 1979, 1991). This capacity for culture evolved as either an exaptation or an adaptation, in that the increased brain size that made it possible had to have been a product of selective forces that favored increased brain size for the advantages it conferred on individuals. It matters not for present purposes whether what was actually being selected for initially was capacity for culture, general intelligence, social intelligence, empathy, cooperation, etc., as they are all likely intertwined in the structure of the brain.

It is also an accepted fact that culture itself has evolved in conjunction with the capacity for it (e.g., see Durham, 1991; Shennan, 2002). That it has evolved is clearly evident in even the most cursory comparison of the earliest archaeological assemblages with more modern ones. It is also generally accepted that culture is an adaptation. That is, socially transmitted learned behaviors are the primary means by which humans interact with the environment for their somatic well-being and reproductive success and that there has been selection over time for increasingly efficient energy-matter transfers in such interactions (e.g., compare early hominin scavenging with modern human animal husbandry). The increased numbers and density of *Homo sapiens* within their habitat range over time and their increased habitat range over that same time frame are an obvious measure of just how successful that adaptation has been. However, in regard to the mechanics of cultural evolution, there has been less of a general consensus.

M. Rosenberg, *The Dynamics of Cultural Evolution*, Studies in Human Ecology and Adaptation 12, https://doi.org/10.1007/978-3-031-04863-0_3

The evolution of culture has been a central concern of Anthropology ever since it began to coalesce as a scientific discipline in the nineteenth century, in loose association with the development of evolutionary theory and characterized by the then-prevailing belief in the biological (i.e., racial) basis for cultural evolution (see Harris, 1968; Degler, 1991 for good summaries). However, while Biology abandoned the evolutionary theories of Lamarck and Herbert Spencer as unworkable at the beginning of the twentieth century, in sole favor of Darwinian evolution, Anthropology only abandoned them as regards human biological evolution. With respect to cultural evolution, it purposefully retained significant Lamarckian elements for two reasons. First, it was recognized that cultural transmission was sufficiently different from biological inheritance that culture was capable of changing in accordance with something akin to Lamarckian principles (e.g., see Kroeber, 1917). Second, by continuing to embrace Lamarckian mechanisms despite their having been thoroughly discredited in Biology, Anthropology was arguably furthering its then-current agenda of emphatically distancing the concept of culture from the concept of race. In other words, not only was it theoretically appropriate for the Boasians and their anthropological successors to not completely abandon Lamarckian mechanics when dealing with culture change, it was programmatically expeditious for them to thoroughly embrace Lamarckian mechanics while implicitly rejecting Darwinian mechanics (and the racialist theories with which they had by then increasingly come to be associated) by simply ignoring them in matters of culture change (see Degler, 1991: 92ff.; see also Dunnell, 1980).[1]

Kroeber (1917) was quite explicit about both his embrace of Lamarckian mechanisms and his denial of Darwinian mechanisms as they applied to his and the other Boasians' historically particularistic view of cultural evolution. Though perhaps less consciously programmatic, such an embrace of Lamarckian mechanisms continued to also characterize the next generation of cultural evolutionists (e.g., Steward, 1955; White, 1959; Sahlins, 1960; Service, 1962) that arose in the mid-twentieth century, with their renewed interest in the broader aspects of sociocultural evolution, as opposed to the historical particularism of the Boasians—i.e., the evolution of culture as opposed to the evolution of cultures. The problem is that Lamarck's evolutionary theories, particularly as adapted by Herbert Spencer to sociocultural evolution and ultimately adopted by the mid-twentieth-century (anthropological) evolutionists, encompassed a good deal more than just concepts concerning a specific set of transformational mechanisms that might perhaps be considered as workable when dealing only with culture. Of these, the one of most immediate concern is the concept of evolution as progress, a holdover from the unilinear evolutionism of the late nineteenth century (see Harris, 1968; Carneiro, 1973; Dunnell, 1980). As noted at the outset of this book, the very term "evolution"—Spencer's favored term and reluctantly accepted by Darwin as synonymous with his more favored "descent with

[1] The continuing anathematic reaction in some quarters to any role for Darwinian mechanics in the evolution of human behaviors has strong roots in a continuing fear of potential racist misuse (e.g., see Irons & Cronk, 2000).

modification"—originally meant a *progressive* unfolding along a *set sequence*. As also noted at the outset, the term "evolution" still retains that original connotation in Astronomy, when, for example, astronomers talk about the evolution of stars. Unfortunately, it also continued to connote that in the various models of sociocultural evolution that came to the fore in the middle of the twentieth century, such as the band-tribe-chiefdom-state sociopolitical evolutionary sequence (e.g., Service, 1962).

The obvious differences in the mechanics of change aside, this more subtle distinction between the essentially non-teleological nature of Darwinian evolution and the progressiveness of Lamarckian evolution also went largely unrecognized by archaeologists, as Processual Archaeology began to move beyond culture history in the post-WWII period to enthusiastically address issues of adaptation and processual evolutionary change (e.g., Binford, 1968). That is, as (particularly American) archaeology came to increasingly see itself as anthropologically oriented, rather than historically oriented, archaeologists implicitly adopted the Lamarckian/Spencerian evolutionary views that dominated the evolutionism of sociocultural anthropology,—not the Darwinian evolutionary views that dominated physical anthropology,—without necessarily being fully aware of all the less obvious distinctions between them. Thus, archaeologists often talked about evolution and adaptation but failed to recognize that their models tended to unconsciously invoke concepts of progress and the astronomical (i.e., Spencerian) view of (social) evolution more so than they did the biological view of evolution (see Dunnell, 1980; Rindos, 1984). So, for example, Binford (1968) was quite content to argue that food production evolved in the Levant at the end of the Pleistocene in the context of population pressure, simply because it was advantageous in the sense that more people required more food. While food production no doubt would have been advantageous under the conditions he postulated, nowhere did he acknowledge that just because something is potentially advantageous does it mean that it will necessarily appear to be selected for (or against). That is, in neo-Darwinian evolution, the appearance of advantageous novelties is not inevitable; the appearance of novelties is driven by processes that operate independently of the selective forces that operate on those novelties, and they must therefore be accounted for separately. Neither did he nor any of his theoretical allies (e.g., Cohen, 1977) postulate a mechanism that might produce the requisite behavioral novelties, implicitly assuming instead that the behavioral novelty would inevitably arise by virtue of its own obvious advantageousness. One can also see a similar focus on functional advantageousness in many of the cotemporary models for the subsequent rise of politically hierarchical societies, particularly those models that viewed their rise in voluntary terms.

In essence, this point was central to Rindos' critique of processualism when he stated that

> The [symbolic] system is, frankly, not a particularly effective one for the creation of 'intelligent' or directed variation. Far too much ambiguity exists within it, and it seems to generate new variants more at the behest of its own (evolved) rules than as an obvious means for the harmonious integration of culture and environment. (Rindos, 1985: 72)

While one can easily argue that Rindos seriously overstated the point and while his critique was aimed at furthering a very different view of how cultural novelties are produced than the one that will be put forward here, the crux of his criticism (see also Dunnell, 1980) is nevertheless valid. Culture does not inevitably "unfold" by Lamarckian/Spencerian transformations to ever better (functionally) adapted/organized states of being relative to then-current environmental/social conditions. In a nonprogressive view of evolution, the processes that produce novelties operate[2] independently of the processes that select from among such novelties.

These critiques of processualism were part of a more widespread trend beginning in the 1970s in some quarters of the social sciences,—notably Psychology and Anthropology/Archaeology, but also Economics—that sought to replace the still-dominant Lamarckian/Spencerian evolutionism of the social sciences with a view of human behavior rooted in Darwinian evolutionary mechanics. The common denominator within this trend was the view that (despite the past racialist/racist excesses of the first half of the twentieth century, justified in large part by recourse to genetics and biological evolution, however the latter was viewed) Darwinian evolutionary mechanics cannot be excluded from any discussion of human culturally patterned behaviors that seeks to truly understand both the evolutionary basis for such behaviors and the evolutionary mechanics that produced them.

Out of this growing general interest in the role of Darwinian mechanisms in the evolution of human behavior, culturally structured or otherwise, several intellectual streams emerged from within Psychology and Anthropology/Archaeology (for comparative discussions, see Blurton Jones, 1990; Durham, 1990; Spencer, 1997; Smith, 2000b; Mesoudi et al., 2006). While it makes for convenient discussion to view each of these intellectual approaches to the evolution of human behavior as theoretically bounded, in practice, many individual researchers did/do not fall clearly into one category or another because they rely on insights derived from some variable combination of these basic approaches. This is possible because, despite some relatively early theoretical skirmishing over whether or not a given approach was studying the evolution of human behavior correctly (e.g., Symons, 1989, 1992 vs. Turke, 1990; Boone & Smith, 1998 vs. Lyman & O'Brien, 1998), it was soon enough recognized that there was often sufficient room for at least some theoretical reconciliation of these diverse approaches (e.g., see Barkow, 1989; Blurton Jones, 1990). The conceptual underpinnings of these various approaches are therefore worth reviewing briefly before proceeding. The aim here is not to critically examine these various approaches' respective strengths and weaknesses. It is simply to lay out the most important borrowings and insights from Biology that have achieved a significant degree of acceptance so that a foundation is laid for what will follow.

However, it is important to note that in keeping with the Neo-Darwinian synthesis that was current at the time it was being imported, the founder's effect was that all

[2]To say that two processes *operate* independently of each other is not to say they must of necessity *be* independent of each other (cf. Walsh, 2015).

these approaches were more often than not reductionist in one manner or another and grounded in the analysis of minimal cultural/behavioral units. More importantly, of those that focused on evolutionary mechanisms, most were either explicitly or implicitly concerned with only the form of selective forces and by default treated the objects of those selective forces as being, for all intents and purposes, passively subject to those forces, in the same manner as genes are.

The Adapted Mind and Behavioral Propensities

While a detailed discussion of the intellectual precipitants within Biology of the changes to the concept of evolution within the social sciences during the 1970s is too tangential to delve into here, one that certainly had an important impact, particularly in Psychology, was Wilson's (1975) popularization of the concept of "Sociobiology." In a nutshell, Sociobiology proposed that social behaviors such as altruism can be evolved, genetically *based* (which is not necessarily to say genetically *determined*) adaptations *even in humans*. It did so without recourse to the then-still seemingly discredited concept of group selection and in direct contradiction to the then-still dominant view in the social sciences that human behavior is for all practical purposes completely flexible, with individuals at birth being more or less a behavioral *tabula rasa* whose behavior is ultimately shaped largely if not solely by culture. In practice, this social science view accorded no role whatsoever to evolved behavioral propensities as regards human behavior, a point increasingly at odds with the data derived from the by-then burgeoning field of primate ethology.

Evolutionary Psychology is perhaps the most direct theoretical descendant of Sociobiology, in a strong sense being an *appropriately modified* 'Human Sociobiology,' which has avoided the Sociobiology name in favor of one that carries less negative semantic baggage. It is concerned with what is often colloquially called 'human nature'—, the biological (i.e., genetic) basis for human behaviors, as rooted in the evolved architecture of the human mind (e.g., see Tooby & Cosmides, 1990, 1992). Put another way, Evolutionary Psychology is interested in the 'games people *tend* to play' and why they tend to play them. However, it is only peripherally interested (if at all) in the precise structure of the playing fields (i.e., culture) on which those games were played out in the past, the learned rules of play, and even less interested in the structure of the playing fields on which those games are played out in the present or how the structure of those playing fields evolved. Thus, of the abovementioned theoretical approaches, it is the one furthest removed from the direct study of culture and cultural evolution, and it expresses the least explicit interest in either culture or cultural evolution. However, that is not to say that Evolutionary Psychology is irrelevant to the study of cultural evolution, since "[c]ulture is not causeless and disembodied" (Cosmides et al., 1992: 3). In other words, Evolutionary Psychology's lack of explicit interest notwithstanding, the innate tendency of people to 'play games' and the "games they tend to play" are not irrelevant to the study of how the culturally structured 'playing fields' on which such games play out have evolved. This is because innate tendencies give us a peek into the black box of the mind and thus intent.

> Of course, the social system is not ... self-ordering due to its own functionally integrated mechanisms. It is more like an ecosystem or an economy whose relationships are structured by feedback processes driven by the dynamic properties of its component parts. In this case, the component parts of the population are individual humans, so any social dynamics must be anchored in models of human psychological architecture. (Tooby & Cosmides, 1992: 47)

As summarized by Symons, the goal of Evolutionary Psychology is to understand "the psychological mechanisms that underpin human ... behavior, and ... the selective forces that shaped those mechanisms" (1992: 137). In other words, it is the study of psychological mechanisms as biological adaptations produced by Darwinian processes. However, they are generally not considered to be adaptations to some set of present conditions because present conditions have not typically been in existence long enough to incrementally produce the adaptations in question by means of Darwinian selective processes operating biologically. Rather, they are considered to be adaptations of the mind to the long, largely unchanging social and physical environmental conditions in which humans existed up until the end of the Pleistocene and the (cultural) evolution at that time of settled village societies based on food production. Typically referred to as the "adapted mind" (e.g., Barkow et al., 1992; Cosmides et al., 1992), these adaptations consist of a set of evolved (and thus genetically based) reductively modular cognitive mechanisms geared to the successful performance of generally defined tasks, typically relating directly or indirectly to reproduction (e.g., male attraction to traits indicating fertility in females), *but also to somatic well-being*, as with cooperation.

The importance of the concept of the "adapted mind" to what will follow is that it proposes that we as humans are prone to biologically selected-for behavioral propensities, which, while biologically based, are not biologically determined and that human behavior is not solely a product of learning—that we are not a behavioral *tabula rasa*. While Evolutionary Psychology is chiefly concerned with only certain classes of behavior, typically those produced by sexual selection, if it can be plausibly demonstrated that other behavioral propensities exist because they were biologically selected for in the past by selective forces, they can arguably be cited as the basis for the ascription of agency.

Conditional Strategies, Phenotypic Plasticity, and Cost-Benefit Calculations

A second important source of biological influence, one affecting mostly Anthropology and Archaeology in the form of an interest in Human Behavioral Ecology, was the rapid development of animal Ethology and Behavioral Ecology in the post-WWII decades. Human Behavioral Ecology is less interested in human nature as an adaptation forged in the past that we carry as psychological baggage into the present than it is in the decision-making capability of the human brain as an adaptation that continues to function as effectively as ever in the present. That is, Human Behavioral Ecology is less concerned with the central tendencies reflective of biologically based behavioral propensities (modular or otherwise) and instead more concerned with the biologically based ability of the brain to calculate the context-specific costs and benefits of potential actions and the context-dependent behavioral variability such calculations generate. In still other words, it is less interested in the behavioral propensities that are part of our nature as a species than they are in the behavioral flexibility that has been the key to our success as a species (and to a lesser degree, the basis for the decision-making capacity that is the basis for that behavioral flexibility).

Specifically, it considers the decision-making capacity of the human brain[3] to be an evolved adaptation that generates behaviors that enhance fitness in the context of immediate environmental conditions by means of what are often called evolved "decision rules" (Krebs, 1978) or "conditional strategies" (Smith, 2000b).[4]

> These are abstract and somewhat metaphorical ways of conceiving the covariation of behavior and socioecological environment. Having the general form: In context X, do α; in context Y, switch to β (Smith, 2000b: 30).

The behavioral variability resulting from the operation of these decision rules is considered an expression of "phenotypic plasticity," defined as "a phenotype's... capacity to respond differentially to varying environmental conditions" (Boone & Smith, 1998: 144). The phenotype in this case is the elaborate decision-making capacity of the human biological brain and its evolved decision rules, the plastic phenomenological expressions of that phenotype being some variety of observable behaviors, each a product of some set of specific environmental conditions. Simply put, the central insight is that we have become adapted to adapt.

What we have here is not Evolutionary Psychology's evolved behavioral tendencies being the biologically based adaptation of interest. Rather, it is the mental, decision-driven capacity to adapt differentially to short-term variability in environmental conditions (by what very much amount to Lamarckian mechanisms) that is considered the biologically based adaptation of interest (e.g., see Smith, 2000b). In other words, the actual Darwinian adaptation in this case is the brain's complex of evolved decision rules that confer the capacity to behaviorally adapt in the short-term to changing conditions. It should be noted that the intentions produced by the decision rules are not to 'adapt' in some idealized Lamarckian sense. Rather, they are to maximize returns on variable immediate goals, such as the gaining of resources at a specific point in time or achieving reproductive success (see Smith, 2000b). In other words, in contrast to Evolutionary Psychology's primary interest in the kinds of games people have a propensity to play, Human Behavioral Ecology is

[3] This decision-making capacity is also central to other approaches to the evolution of culture (e.g., Dual Inheritance). However, they do not focus specifically or exclusively on decision-making as a product of evolved decision rules.

[4] Behavioral Ecology is 'agnostic' about the biopsychological mechanisms underlying the decision rules (see Smith, 2000b: 37).

primarily concerned with how people strategize their game playing. Thus, as in the study of economics, games theory (e.g., Maynard Smith, 1982) is often profitably employed analytically.

The study of conditional strategies is reductionist in its approach, in that it conducts the study of decision rules by means of a "*piecemeal approach*" (Smith, 2000b: 29, emphasis in original) that focuses on the individual behavioral patterns affected by some particular 'rule' and on which selection presently acts in classic Darwinian terms. The stated rationale for this piecemeal approach is the practical dictum that "complex socioecological phenomena are [more] fruitfully studied piece by piece—in reductionist rather than holistic fashion" (Smith, 2000b: 29). Thus, this reductionism is not necessitated by the theoretical premise that higher levels of organization must be reducible to some number of lower-order units (for how else could it be simple enough to be genetically coded); rather, it is merely a methodological convenience.

Like Evolutionary Psychology, Human Behavioral Ecology is primarily interested in demonstrating that certain behaviors are adaptive, not so much in how they evolved. One consequence of this focus and Behavioral Ecology's methodological reductionism is that the influence of culture on decision-making tends to be finessed. That is, humans are social animals *par excellence*, and the immediate environmental conditions in which individuals make decisions obviously include social and cultural variables. However, for Human Behavioral Ecology, culture generally tends not to be factored into the environmental picture except in the most basic terms and then only to the degree absolutely necessary, presumably on grounds that it is unnecessary clutter that potentially obscures the economic/reproductive pattern of interest. Thus, with a few exceptions (e.g., Boone, 1992), Behavioral Ecology as a research focus is not particularly concerned with the evolution of specific cultural phenomena on any but the most abstract level (e.g., see Boone & Smith, 1998).

Lastly, it should be noted that the decision rules that interest Behavioral Ecology can operate to optimize the somatic well-being of individuals (e.g., Smith, 1988) as well as enhance their reproductive success (e.g., Borgerhoff-Mulder, 1990). Thus, Behavioral Ecology is interested in the operation of behavioral adaptations that enhance the fitness of individuals with respect to not just sexual selection, but natural selection as well. It is the proposed existence of such decision rules and the general ability of humans to calculate the costs-benefits of alternative courses of action that most bears on what will follow.

Cultural Selection and Cultural Drift

The concepts of cultural drift and cultural selection work their way into the mechanics of cultural evolution most importantly via Cavalli-Sforza and Feldman (1981) and influence the development of both what came to be called Evolutionary Archaeology (e.g., Dunnell, 1986, 1989; Rindos, 1984, 1985) and Dual Inheritance Theory (Durham, 1979, 1990; Boyd & Richerson, 1985). Unlike Evolutionary Psychology and Human Behavioral Ecology, which are both primarily concerned with behavioral tendencies directly rooted in one way or another in the biological structure of the brain and largely isolated from anything beyond passing references to cultural systems, both of these approaches were, from the very beginning, concerned with the evolution of culture, albeit in very different ways.

Evolutionary Archaeology is specifically concerned with the evolution of the material products of culturally based behaviors, as recovered from the archaeological record. Central to this approach is the concept of the "extended phenotype," taken from Dawkins (1982). Simply summarized, the extended phenotype refers to the full range of (however indirect) phenomenological expressions of some genetic instruction set. Examples of such expressions would include a hermit crab's appropriated shell, a bowerbird's bower, not to mention the bower's decorative trinkets, and a beaver's dam. That is, each of these is a product of behaviors that characterize the respective species as a whole and must therefore have a basis in that species' basic genotype, even though it is not a direct biological aspect of that genotype's phenotype. Evolutionary Archaeology considers the material manifestations of culture (i.e., artifacts) to be expressions of the extended human phenotype (see O'Brien & Lyman, 2000: 8–9). Since selective forces operate on phenotypes, they perforce also operate on the artifactual products of culturally patterned behaviors on the basis of their functionality, as well as through them on their users by reference to the behaviors represented by the artifacts themselves. Thus, Evolutionary Archaeology is concerned with cultural artifacts (and the behaviors that they are proxies for) as evolving adaptations in their own right by virtue of selective forces operating on their traits (e.g., see Dunnell, 1980; Lyman & O'Brien, 1998; O'Brien & Lyman, 2000). Its focus is on documenting the evolution of such cultural adaptations using the archaeological record, much like Paleobiology documenting the evolution of biological species using the fossil record (see Lyman & O'Brien, 1998).

In contrast, Dual Inheritance Theory is primarily concerned with culture as a "system of knowledge"(Durham, 1991: 7) that has coevolved with the biological capacity for culture and that "a similar causal relationship between ecological and evolutionary processes governs cultural evolution" as governs biological evolution (Boyd & Richerson, 1985: 291). In this view, culture and culturally based behaviors stand in the same evolutionary replicator-vehicle relationship to each other as do genes and the organisms that carry them in the biological domain (e.g., see Cavalli-Sforza & Feldman, 1981; Boyd & Richerson, 1985; see also Durham, 1979, 1990, 1991; Bettinger, 1991; and Shennan, 2000, 2002). That means that Dual Inheritance Theory explicitly defines its object of study—culture—in ideational terms instead of materialistic terms. In other words, culture is viewed as a separately evolving information system (from the genetic system) that generates phenomenological behaviors, which are subject to biological forms of selection. These selective forces feed back differentially into the transmission/selection (and hence evolution) of the information system that they are the phenomenological expressions of.

It is an accepted fact in Anthropology that culturally transmitted (i.e., socially learned) behaviors are variable, as are the artifacts produced by such learned behaviors, if for no other reason than the differing skill levels of individuals and their differing understandings as to how some specific thing should be done (e.g., see Goodenough, 1981). Given variability, it is inevitable that selective forces will operate on that variability, whether it is based on functionality, aesthetic appeal, or any one of many other possible conscious or unconscious criteria.

Both Evolutionary Archaeology and Dual Inheritance Theory propose that there are two different modes of selection operating on the evolution of culture, as defined by each. These are Darwinian selection operating on the biological fitness of individual culture bearers by reference to their culturally structured behaviors and cultural selection affecting the transmission between individual culture bearers of the actual elements of the information system by reference to the relative success of individuals engaging in specific behaviors or using specific variants of artifacts. Rindos (1985) termed the former cultural selection of the first type or CS1 and defined it as "*the direct selective advantage given individuals because they possess a specific cultural trait*" (1985: 73, emphasis in original). Thus, CS1 refers to selection *for* variants of cultural phenomena operating on biological individuals with specific reference to their cultural traits, and this is not really what has since come to be understood as "cultural selection," being instead natural (or even theoretically sexual) selection operating on individuals by reference to the variable expressions of the extended phenotype. He defined the second, which he termed cultural selection of the second type or CS2, as "the factors that bring about the adoption or nonadoption of a particular symbol or trait *without concern for* its effect on the [biological] phenotype" (Rindos, 1985: 73, emphasis added). Thus, in a manner of speaking, CS2 refers to selection *by* culture, and this is what has specifically come to be considered cultural selection,—the culturally influenced human choices affecting the differential replication (i.e., transmission) of cultural traits (e.g., see Boyd & Richerson, 1985).

While Evolutionary Archaeology recognizes the existence of both natural and cultural selection, it is primarily concerned with the former,—for the most part natural selection operating on individuals with reference to cultural adaptations as expressed in artifacts. Cultural selection is implicitly assumed to be operating at some level, as is drift, to produce variability in the extended cultural phenotype, thus making such variability available for further selection on their cultural carriers by natural (and theoretically sexual) selection, but the primary focus is on demonstrating the functional advantageousness of specific artifactual adaptations.[5] That is, to the degree that such variability is functionally as opposed to just stylistically variable, it will be subject to selective forces as opposed to drift—differential transmission based on the degree to which individual varieties of the traits in question affect the efficacy of the artifact and thereby the somatic well-being of their users/makers (Dunnell, 1978; O'Brien & Lyman, 2000). That differential transmission will be reflected in the archaeological record as the changing relative

[5]It was the initial failure, since corrected (e.g., Lyman & O'Brien, 1998), of Evolutionary Archaeology to formally address the differing foci and mechanics of cultural and natural selection that permitted the theoretical slippage criticized by Boone and Smith (1998).

frequencies of traits over time, as the more adaptive artifactual traits are selected for culturally (e.g., see Lyman & O'Brien, 1998) by virtue of their success at enhancing the functionality of the artifacts they are elements of, and through the artifacts the well-being of the artifacts' makers, such that they are seen by others as being relatively more successful and therefore emulated in how others make/do things.

Dual Inheritance Theory, on the other hand, focuses primarily on cultural selection and its role in the evolution of culture (e.g., Boyd & Richerson, 1985; Durham, 1991). The justification for this focus on cultural selection is that natural selection only loosely influences cultural transmission, which is also selective, but not *necessarily* geared toward maximizing biological fitness at the level of individual culture bearer (see Boyd & Richerson, 1985; Durham, 1991). Thus, the emphasis is on understanding the mechanics of cultural evolution as the fundamental first step to determining why a given cultural complex evolved because we cannot a priori assume it to have evolved to maximize the *biological* fitness of the individuals practicing it.

Darwinian selection and cultural selection are fundamentally different from each other. Cultural transmission is neither rooted in biological reproduction nor is it rooted in any other kind of physical process. Thus, it is not constrained by the same physically mechanistic vertical (i.e., parent-child) transmission processes that govern transmission of the genetic code in multicellular organisms. Culture is transmitted through the observation by others of the behaviors it generates in individuals, but ultimately it is not a behavioral system (see Goodenough, 1981). It is, therefore, potentially Lamarckian in its operation.

> Like many other organisms, humans adjust their phenotypes in response to their environments through learning and rational calculation. Unlike most other organisms, humans can culturally transmit the phenotypes so acquired to the next generation. Thus ordinary learning in combination with cultural transmission acts to create a kind of "Lamarckian" effect. (Boyd & Richerson, 1985: 9)

Cultural transmission is a mental process based on either observation or instruction, and while vertical transmission plays a very important role (e.g., Hewlett & Cavalli-Sforza, 1986; Shennan & Steele, 1999; see also Shennan, 2000: 813), the Lamarckian aspects also operate to make possible horizontal and oblique transmission, which can bias the transmission process in several ways to favor some cultural variants over others. Such biases arise from the perception on the part of individuals that *deciding* to adopt one possible variant over another is more likely to lead to a more favorable outcome (see Boyd & Richerson, 1985: 10,94ff.; see also Shennan, 2002: 50).

The potential disconnect between cultural selection and biological fitness arises from the fact that 'favorable' outcome is determined by an individual's broader set of culturally framed understandings regarding the desirability of possible alternate outcomes. When the favorable outcome is something as basic as successfully getting food and cause-effect is both directly perceptible and reliably ascertainable with a reasonably high degree of factual accuracy, then the selective biases may very well favor the cultural variant that most enhances biological fitness, but even then not necessarily (see Boyd & Richerson, 1992: 75ff). However, when it is something as complexly nebulous as gaining social approval, the 'favored' outcome may enhance *net* biological fitness—as in: enhance social chances for attracting most/best mates at the expense of sharing food or other altruistic acts—or it may not enhance net biological fitness at all. And the degree to which it may not enhance net biological fitness is directly related to the complexity of the cultural overlay and the resulting necessity of weighting primary, secondary, tertiary, etc., costs-benefits against one another, as in weighing the primary benefit of eating an animal versus the higher-order benefit of gaining supernatural favor by sacrificing it. This is compounded by an individual's difficulty in accurately discerning the true somatic cost-benefits of progressively higher-order culturally framed goals, not to mention a given individual's skill at accurately calculating.

As noted above, Dual Inheritance Theory also considers Lamarckian type processes to play a role in the generation of cultural novelties. Thus, directed or guided variation is also possible and rooted in the same human mental capacities invoked by Human Behavioral Ecology.

> Adaptation through rational calculation proceeds by the collection of information about the environment, the estimation of the results of various alternative patterns of behavior, and the evaluation of the desirability of the alternative outcomes.... It is these guiding criteria that translate variation in the environment into a directional, *often* adaptive, change in phenotype, which is then culturally transmitted to subsequent generations. (Boyd & Richerson, 1985: 9, emphasis added)

All this is not to say the resultant novelties will invariably be adaptive, much less optimally so; it is just to say that their generation will tend to be problem directed and goal oriented, i.e., purposive, and that they therefore have a greater probability of being adaptive relative to the environmental conditions that sparked them than would be randomly generated novelties. To put it another way, if the problem is finding a mode of transportation to cross bodies of water, the novelties tried are likely to more closely resemble what we commonly call boats (by virtue of having the ability to float on water) than what we commonly call vehicles (by virtue of having wheels).

The problem with Lamarckian processes and agency, as Evolutionary Archaeology sees it, is that intent on the part of the no-longer living is not knowable scientifically and therefore cannot be reliably distinguished from drift (i.e., random change) as the source of archaeologically recovered novelties (e.g., Rindos, 1985). Thus, Evolutionary Archaeology (like Paleontology) satisfies itself with the explanation that any novelties necessary for some selective outcome existed simply because they had to have existed in order to produce the observed selective outcome (e.g., Dunnell, 1980: 62; Rindos, 1984, 1985; Lyman & O'Brien, 1998). The danger lurking here is the possibility of a descent into adaptationism (see Gould & Lewontin, 1979). More importantly, the premise that intent is always not knowable is debatable, a point that I will return to below.

Systems, Hierarchy, and Multilevel Selection

From the beginning, Dual Inheritance Theory viewed culture in both systemic and hierarchical terms (e.g., Boyd & Richerson, 1985, 1987; Shennan, 2002: 91). That is, the focus was not just on the mechanics of transmission as regards the information system but also on culture as a shared "population-level phenomenon" (Boyd & Richerson, 1985: 6), characterized by emergent group-level evolutionary dynamics.[6] Examples of such proposed between-group dynamics are "imposition" and "opposition" (Durham, 1991), which together address the question of how evolution can produce the exploitation of one constituent group by another in complex societies.

Imposition, simply put, is the "preservation [of some variant] by compliance with the decisions of others" (1991: 198) regardless of the imposed-on group's interests and/or wishes. Imposition is a consequence of competition, and imposed variants are more often than not a variant justified by the dominant information system of the winners of that competition and presumably often enough contested to at least some degree by the losers of that competition on whom it is imposed. Opposition is often intertwined with imposition. In a nutshell, opposition is the preservation of some variant that specifically acts to the somatic and/or reproductive detriment of individuals in groups characterized by that variant (1991: 368ff). Leaving aside for now the occasional theoretical slippage between ideational and behavioral systems, as regards just what kind of higher-order entities these dynamics refer to, a key point, one that I will return to in more detail later, is that cultural selection has the capacity to overpower biological modes of selection. The more immediate point, however, is that oppositional variants can be imposed *by one group-level entity on another.* Thus, this hierarchical view of cultures and societies as forms of higher-level entities—"superorganisms," so to speak (Richerson & Boyd, 1999)—allows for discussion of group/multilevel selection (Boyd & Richerson, 1987, 1990; Soltis et al., 1995) independently of the classic debate centered on altruism and much more in the mold of the theoretical fallout from evolution proceeding as a process of punctuated equilibria.

In any case, sorting out the issue of entities in the cultural and social domains is essential for effectively addressing multilevel selection, macroevolution, and other group-level culturally structured phenomena. For example, Durham's (1991: 371) concept of imposed opposition encompasses, among other things, culturally structured group-level asymmetrical symbiosis and the potential for at times outright socially structured parasitism. This ultimately requires functional working definitions of the "kinds" of social entities involved, as does Soltis et al.'s (1995) foray into cultural group selection. That is, there is a need to systematically relate information

[6]It is worth noting that despite Evolutionary Archaeology's emphasis on traits and its view of higher-order entities as both temporally and spatially fleeting ephemera, traits can be a feature—emergent or otherwise—of such higher-order entities whether transient or not. For example, the traits discussed by Maxwell (1995) are arguably the traits of community-level artifacts produced by cooperative actions.

system to both phenomenological behavioral product and social carrier and biological beneficiary/loser at levels between the individual culture bearer at one extreme and vaguely defined 'cultures,' 'populations' or 'societies' at the other. This is because not all definable kinds of sub-entities necessarily carry/contain the same information system as the larger kind of entity they may be components of—for if they did, there would be no contesting of it—nor do all definable groups engage in ecological interactions of the type that would be affected by natural selection or necessarily feed back into cultural selection through it. Thus, while glossing over these distinctions may suffice in the case of small-scale cultural systems wherein the boundaries of language, culture, society, community, etc., are all coterminus (cf. Goodenough, 1981), it breaks down in larger, more complex social/ideational systems (e.g., classical era Greece → Athens/Sparta → Helots/free Spartans → Perioeci/Spartiates) and requires approaching the problem from a clearly defined taxic perspective.

Punctuated Equilibria

The archaeological record, like the paleontological one, abounds with discontinuities in that the transition from one named archaeological complex to its immediate successor or neighboring offshoot was both archaeologically "instantaneous" (i.e., occurring within less than a century, typically *much* less) and clearly involved at least a major if not a radical cultural reorganization. In fact, it is probably not excessive to make the claim that such discontinuity characterizes the emergence of virtually every distinctly named complex in the archaeological record. Thus, it is not surprising that cultural evolution in the form of punctuated equilibria has at times been invoked by individuals working in a variety of theoretical frameworks (not all of them explicitly Darwinian) to explain these discontinuities.

For example, Renfrew (1978) viewed punctuational episodes in terms of tipping points, and Dunnell (1980: 59–60) acknowledged that in general terms punctuated equilibria as well as incremental processes may both characterize the evolution of culture, though of theoretical necessity Dunnell was focusing on the tempo and not the mode of change. More specifically, examples of claimed punctuational change (with or without intervening stasis) have been made with respect to the pattern underlying the emergence of state systems in general (see Diener, 1980; Spencer, 1987, 1990; Yoffee, 1991), the emergence of named cultural complexes in the Levantine Neolithic (e.g., Bar-Yosef & Belfer-Cohen, 1992; Rosenberg, 1994), the American southwest (e.g., Berry, 1982), Mesoamerica (Spencer, 1990, 2009), and the North American Pacific Northwest (Prentiss & Chatters, 2003; Chatters, 2009; Prentiss, 2009).

As with Dunnell's, some such claims (e.g., Renfrew, 1978; Yoffee, 1991; Bar-Yosef & Belfer-Cohen, 1992) were based on the conservative understanding of punctuational change as characterized simply by its pace and not also by distinct evolutionary processes. However, some others also tried to address the distinct processes that produced stasis and the processes that break such stasis by producing systemic disintegration (Diener, 1980; Spencer, 1990, 1998; Rosenberg, 1994; Prentiss & Chatters, 2003), as well as related macroevolutionary processes (e.g., Prentiss et al., 2009: passim). Much of this work addressing the mechanics of macroevolution was done by reference to the archaeological record and under the rubric of a resurrected processualism, sometimes labeled Evolutionary Processualism.

As in Paleontology, there is substantial evidence for a pattern of punctuated equilibria in the archaeological record. Thus, the critical question is what are the environmental, social, etc., processes that produce punctuated equilibria in place of incremental change. Whether the *biological* processes that produce punctuated equilibria are agreed upon or not is surely important to biologists. However, that ongoing biological debate is extraneous to the evolution of culture. This is because diverse aspects of ethnological theory already arguably contain the theoretical explanation for a tendency to stasis. They arguably also contain mechanisms that can rapidly produce radically *new* systems independently of selection and other mechanisms capable of producing "cascading effects thereafter." In other words, the mechanisms arguably capable of producing punctuated equilibria in the evolution of culture are already widely (though not uniformly or universally) accepted as existing. As in biology, grappling with punctuated equilibria in the matter of cultural evolution again requires approaching the problem from a clearly defined taxic perspective.

Niche Construction

Niche construction theory is the most recent import from Biology to gain a degree of acceptance (e.g., Zeder, 2012, 2016), facilitated by its consilience with many of the above insights. As noted previously, the core concept here is that organisms and environments coevolve. Organisms, by engaging in life activities, change their environments and the selective regimes that are a feature of those environments, while also concurrently becoming adapted to those changed environments as a result of any changes to the selective regime produced by their life activities. Its relevance to the evolution of culture was recognized from the very beginning by Odling-Smee et al. (2003).

> Genetic processes, ontogenetic processes, and cultural processes operate at three distinct but interconnected levels...: that is, learning is informed, but only loosely, by genetic information, and cultural transmission may be informed, but not completely specified, by both genetic and developmental processes. Genes may affect information gain at the ontogenetic level. In addition, ontogenetic processes, particularly learning, may be affected by cultural processes, while population-genetic processes may be affected by both ontogenetic processes and cultural processes when humans modify their selective pressures. (2003: 254)

It is particularly important to note that just as Evolutionary Psychology invokes behavioral propensities, in most cases largely produced by sexual selection, the above quote, by the use of "informed ... by genetic information," also clearly invokes behavioral propensities in niche construction. Given its roots in Evolutionary Ecology, Niche Construction Theory's primary applications thus far as regards the evolution of culture have essentially focused on the evolution of economic adaptations (e.g., Zeder, 2012, 2016), much as did a more primitive version of such economic coevolution integral to Rindos' (1984) model for the evolution of agriculture, produced more or less under the rubric of Evolutionary Archaeology. Thus, the behavioral propensities being invoked are economic in nature and for that reason both a product of natural selection and available to be further operated on by it. The lingering question is as follows: What are those propensities? More importantly for present purposes, can they be reliably invoked as the basis for agential actions?

Agency in Evolution

Agency, however abstractly, was always integral to processualism, and that has carried over into its current, properly evolutionary expression as Evolutionary Processualism (Spencer, 1993; Prentiss et al., 2009). It is also abstractly integral to most of the other imported concepts. I say "abstractly" because their focus has tended and still tends to be more on the mechanics of cultural evolution—i.e., the evolution of culture—than on the evolution of specific cultural systems, which would require explicit reference to agency. That is, whatever their specific explanatory focus, what all the other importations from biology focusing on evolutionary processes have in common is a focus on some mode of selection while for all *practical* purposes holding the objects of those selective forces as passive reactors to those forces that will go wherever selective forces take them. Such selection can be just in the form of cultural selection in the transmission of the information system, biologically based selection operating on behaviors and affecting either the reproductive or somatic well-being of the actors, or it can be on the biological effect affecting the (culturally) selective transmission of those behaviors, whether the behaviors are biologically or culturally based. The role of agency is avoided either because it simply is not seen as important enough due to its presumed obviousness or knowable enough to devote attention to. That is, the key to understanding evolution is seen to be understanding selection because selection is what produces outcomes (e.g., Jones et al., 1995). Thus, while some have pointed to the fact that people behave purposively (e.g., Cronk, 2006; Pagel, 2006), at best it is abstractly integrated into niche construction (e.g., Mesoudi et al., 2006), and motivation is still simply taken as a given much the same way that mutation tends to be in biological evolution as seen from the paleontological record—it happened; nothing much to see here; let us move along.

The problem with selection-focused explanations is not that they are intrinsically wrong; it is that, as noted before (Rosenberg, 1990: 405; see also Bettinger, 2009), they are at best incomplete in some way or another by any *anthropological* yardstick. Novel cultural expressions are typically invoked in a cultural systemic vacuum, with no structural or theoretical consideration given to possible biases affecting what behavioral complexes can or cannot be said to exist for selection to operate on under any given set of conditions (but see Jones et al., 1995: 18 and Lyman & O'Brien, 1998 for minor exceptions). As Boone and Smith (1998: 152) noted with reference to a wider and somewhat different set of issues, this begs the (not-so-anthropologically-meaningless) question of why some variant would exist to be selected for because a given human behavior is not inevitable under any given set of environmental and cultural conditions. To presume otherwise is teleological, whatever the *telos*. Thus, when there are perfectly valid structural reasons for concluding that some specific behavioral variant should not exist under the cited circumstances, it is necessary to address its proximate causes, specifically purposeful actions, which might produce and perpetuate the novelty said to have been selected for *despite* any cultural systemic forces working to oppose its very appearance, much less its selection.

It is axiomatic that culture is systemic, a complex *whole*; that cultural adaptations ultimately must be materially/behaviorally expressed in order for there to be something for selective forces to act on; and that humans are generally deliberative (but not all-knowing) and purposeful in their actions to a significant degree. The most compelling argument against invoking agency as proximate cause (i.e., motivation) was put forward by Evolutionary Archaeology (e.g., Rindos, 1985). It is that the motivation behind any possible agential actions leading to the production of heritable cultural novelties, being mental, is said to be unknowable scientifically,—*meaning materialistically/archaeologically*. However, even if we momentarily concede the point concerning archaeology's inability to *directly* apprehend proximate causation, that is not the same as saying that proximate causation is not knowable scientifically *at all* (e.g., anthropologically, ethologically, psychologically) and testable archaeologically. Physicists have never directly observed a quark either, nor astronomers directly observed a black hole. But their existence, properties, and effects have nevertheless been indirectly adduced from other phenomena that can be directly observed. More to the point, behavioral propensities, known from studies in other disciplines, will generate patterns at levels above the individual that can be directly apprehended archaeologically because more often than not archaeology deals with aggregates.

In the Modern Synthesis that was the basis for the initial importations, the evolutionary process(es) involved in the production of biological novelties must be completely independent of the forces that select from among them, and any proper evolutionary explanation must reference both processes. Biological novelties are held to be generated mechanistically and independently of selective pressures and thus typically in a truly random fashion with respect to selective pressures (but see Walsh, 2015). Such true randomness means there is no further explanation usually possible for their appearance, nor is a further explanation necessary. However, behavioral novelties are neither produced mechanistically nor necessarily in a truly random fashion. Thus, even though the process of novelty production, being a product of agential action potentially generated by the same selective forces that will also act on the range of actions, is not fully independent of selective forces, it does *operate* independently from the processes of selection. That is, while the production of cultural novelties is not necessarily truly random, it is nevertheless still subject to chance, though such chance is often biased in favor of certain behavioral propensities in the same manner as loaded dice are biased in favor of certain numbers and selection will then operate on all the "numbers" rolled whatever their source. The point is that paleobiologists can safely take the randomly generated and mechanically produced presence of some biological novelty at some point in time as a given. This is simply because it obviously had to have been present to produce a selective biological outcome and, more importantly, nothing further can (or need) be added to that. However, the same cannot be done with respect to cultural evolution with the expectation to produce meaningful *anthropological* explanations that go beyond the evolutionary culture history of isolated phenomena. This is because the novel behavior in question was not mechanistically produced by/in individuals devoid of agency, making its presence require an agential explanation, and that explanation is invariably embedded in a cultural system carried by people, who correctly or incorrectly act purposefully.

Moving Forward

Leaving the issue of agency and proximate causation momentarily aside, there is still the unresolved matter of entities and whether any beyond the most reductive level can be said to truly exist; if so, what they can be said to be; and if so, where they fit in the context of evolutionary processes. The very real problem here is how can one effectively grapple with group selection and other macroevolutionary phenomena in other than the most general terms (e.g., Richerson & Boyd, 1999) without tackling the 'entities issue' and defining the kinds of entities in question. Unfortunately, there has been a general failure to systematically or even consistently distinguish between various levels of the information system (cultural entities as analogs of 'genotype,' 'gene pool,' etc.), the behavioral complexes produced by the various levels of information systems and which are visible to the various selective forces (culture as analogs of 'phenotype,' etc.'), and the biological, typically social carriers of the information systems subject to natural and sexual selection (individual, family, community, society, etc.). The result is occasional conceptual slippage as concerns selective forces between the information systems, the behavioral complexes (e.g., adaptations) produced by the selective employment of that information, and the people who carry the information systems and actually engage in the behaviors.

This is all intertwined with the issue of what exactly is culture and whether it should properly be considered a behavioral or ideational phenomenon. That is, adaptation is a phenomenological product of evolution best studied from a materialist perspective that focuses on behavior, but it is not evolution. On the other hand, *a* culture is not an adaptation. As per Dual Inheritance Theory's informational definition of culture, it is an information system visible to selection only through its phenomenological expression as behavior by individuals and social groups. I briefly touched on this issue elsewhere (Rosenberg, 1994), but a full presentation was not practical in the context of that essay's aims, and more importantly that presentation was flawed because I did not yet recognize the full complexity of the entity (and hierarchies) problem.

Resolving the entity issue requires two things. The first is distinguishing between reproductive/evolutionary and economic/ecological entities, and Eldredge's concept of dual hierarchies arguably offers a promising avenue for doing so. The second is that a taxic perspective—establishing the existential reality of higher-order entities of the kinds said to be involved in the evolution of culture—requires circumvention of Dunnell's "materialist paradox."

> In a materialist view [of evolution], everything is in the process of becoming – the relation we conceptualize as change. .... Things, kinds, and states are artefacts of observation. ... On the other hand, application of evolutionary theory... requires units, things on which selection can act, ... that vary, evolve, or are adapted. (Dunnell, 1995: 34)

Such a circumvention can only be accomplished by establishing a clearly defined network of cultural entities and demonstrating that such entities are not in a constant process of becoming, that is, that such cultural entities are not artifacts of observation. Instead, it must be demonstrated that cultural entities above some reductive level come into discreet being, exist for some length of time, and then cease to exist; and though they may give rise to what replaces them, they do not incrementally change into what replaces them.

Chapter 4
That Complex Whole: The Structure of Culture and the Kinds of Cultural and Culturally Structured Entities

"We often fail to solve our problems because we cannot even identify them. Under such circumstances, conceptual investigations do more than just help. They are the only way out."—Michael Ghiselin

The Problem of Cultural Entities

While attempts to define the concept of Culture abound in the anthropological literature (e.g., Kroeber & Kluckhohn, 1952), the opposite is the case with the categorically messier problem of defining actual cultural entities, with their nested, seemingly endless sub- and macro-sets (but see Goodenough, 1981) of inconsistent size, not to mention seemingly vague vertical and horizontal boundaries at every level. However, in dealing with the evolution of cultures and cultural adaptations, the issue of entities cannot be avoided because it is axiomatic that in order to deal with how something evolves, it must first be determined what that "something" is and whether it can be said to actually be a discreet thing. Nevertheless, the fact of the matter is that significantly more attention has been paid to the mechanics of cultural evolution or to detailing specific culturally structured adaptations than has been paid to gaining an understanding of exactly what kind of entities there are that actually evolve and adapt or become adapted.[1] While this lack of attention is quite understandable in those quarters that do not hold for the existence of evolutionary entities beyond some reductive level, whatever they are said to be, grappling with the problem of entities is an unavoidable necessity for those that do attempt to deal with higher-order entities or logically should (e.g., Boyd & Richerson, 1985; Durham, 1991; Rosenberg, 1994; Prentiss et al., 2009).

[1]The difference between adapt and become adapted depends on whether one is referring to the operation of Lamarckian or Darwinian processes. In Lamarckian processes, entities adapt to their environment by changing in some way that benefits them somatically or reproductively under the given circumstances. In Darwinian processes, they become changed through the differential reproduction of traits that enhance fitness in the context of that environment.

M. Rosenberg, *The Dynamics of Cultural Evolution*, Studies in Human Ecology and Adaptation 12, https://doi.org/10.1007/978-3-031-04863-0_4

As noted by Roscoe (2002), this focus on the evolution of culture in general terms, to the exclusion of cultural systematics – specifics concerning the nature and structural relationship of cultural entities, both hinders our understanding of cultural evolution and has created confusion regarding evolutionary and adaptive units that periodically fueled spurious disagreements. It is as if anthropologists are in the position of biologists attempting to explain the biological evolution of life on earth by explicit reference to only the individual organism at one extreme and phylum at the other, with intermediate entities such as species remaining poorly defined and almost studiously avoided whenever possible precisely because they are poorly defined. At the same time they are often enough invoked using a variety of poorly defined and haphazardly applied intermediate terms (society, culture, community, etc.) when it becomes absolutely unavoidable, which sows confusion (see Roscoe, 2002: 110). This is because the inconsistent usages foster the mistaken impression that the various evolutionary approaches are each, if not fully describing the proverbial elephant,[2] the evolution of culture, at least talking about the same part of its anatomy. This makes one or the other "wrong," when in fact they are in reality often enough dealing with different parts of that anatomy, which remains wholly undescribed.

In the absence of clearly definable culturally related entities beyond the individual culture bearer that can be said to evolve and adapt/become adapted, some approaches, notably Evolutionary Archaeology, Human Behavioral Ecology, and Evolutionary Ecology, have been able to default in the absence of a viable alternative to one form or another of reductionism that falls short in one capacity or another. Others, notably Dual Inheritance Theory and Evolutionary Processualism, have been prone to the conflation of disparate entities under the few available categorical labels, making it difficult to distinguish between genuine and spurious points of disagreement.

In general, there are two broad areas of disagreement in the matter of cultural entities. First, there is the issue of whether cultural entities should properly be defined materialistically or ideationally, and if materialistically, how. Second, there is disagreement over whether or not culturally based entities larger than the individual culture bearer or his/her cultural proxy (in the form of an artifact) can even be properly said to exist. These disagreements are neither new to Evolutionary Culture Theory (e.g., see White, 1975; Harris, 1968, 1979, 1994, 1999) nor are they confined to it within Anthropology (e.g., see Brumann, 1999; Boggs, 2004; Hanson, 2004). The first issue ultimately depends on what kind of entity one is referring to. That in turn requires us to sort out the difference between evolutionary and adaptive, as well as biological and cultural units, and it is here that Eldredge's approach to the species problem using dual hierarchies offers a starting point. The second issue requires us to demonstrate that any such defined entities are discreet

[2]This refers to an Indian fable (also found in the Buddhist sutras), wherein some number of blind men debate the nature of an elephant, based on their limited experience of it from having only touched just one part of it, while being unaware of the larger whole due to their lack of sight.

entities and bounded, that they come into being bounded in some manner from other entities, exist for some length of time, and then cease to exist.

The purpose here is not to produce a concrete application of a taxic approach to some set of real world entities; that seems a rather pointless exercise for present purposes, not to mention being a morass of potentially monumental proportions. Rather, it is just to comprehensively and integratively systematize the various different kinds of entities that evolutionists are dealing with in their various approaches to the evolution of culture. In other words, it is to establish the basis for a taxic perspective as regards cultural evolution. The immediate purposes are to (1) sort out the "entity issue"; (2) lay out a coherent overarching theoretical framework for bridging the insights embedded in too often only *seemingly* conflicting evolutionary views by dispelling the terminological confusion regarding analytical units that fosters spurious conflicts; and (3) provide a tool for evaluating the degree of consilience that the individual theoretical approaches (and the specific explanations they offer) exhibit with the wider (evolutionary) theoretical universe in which they must necessarily be integral. This is because it is only through such consilience that the theoretical synthesis that so many evolutionists (e.g., Smith, 2000b; O'Brien et al., 2003) of different stripes desire can eventually be fully achieved. The ultimate purpose is to lay out the mechanics of cultural evolution in order to elucidate the role of agency in cultural evolution, why agency cannot be ignored, how it can be understood, and, by way of examples, how it has operated to produce both the beginnings of settled village life and ultimately complex societies.

The sticking point here is that using culture to refer to the information used to produce behavior, the behavior produced by the use of that information, and the organic entities that carry the information as well as engage in the behaviors promotes the conflation of disparate things and produces a much more complex cultural facsimile of the species problem than the species problem itself. This is because just as information and behavior are different from each other, so are both different from the group-level social relations based on the information system and structured by it (see Geertz, 1973; Goodenough, 1981; Durham, 1990: 192).

The solution, as with the species problem, is to approach them as elements of different coequal hierarchies of entities. However, in this case, the solution does not involve just two hierarchies, as I attempted to use previously (Rosenberg, 1994); it involves four. Two are organic hierarchies of the type that Eldredge uses to distinguish genealogical (i.e., reproductive) and economic biological entities, and another two are metaorganic hierarchies that distinguish between the informational and behavioral entities that are commonly lumped under the heading of culture.

Human Beings and the Entities of the Organic Hierarchies

Human are biological organisms, subject to natural and sexual selection, and their social systems are culturally structured and hierarchically nested at varying levels of social scale. Eldredge (2003) recognized the need to address human sociality (and by implication culture in general) in the context of his organic hierarchies, but he left off by leaving it as an unexplored third "black box" situated outside his two hierarchies. We now need to look inside that black box and place human beings in the system of dual organic hierarchies proposed by Eldredge. However, while it may be productive to speak of human demes and populations when dealing with humans in purely biological terms, there is no culturally structured social frame of reference inherent in those terms. Thus, more culturally meaningful synonymous terms need to be substituted. In the matter of substitute terms for deme in the genealogical hierarchy and populations in the economic hierarchy, I will here use the term "tribe" in lieu of deme and the term "society" in lieu of population.

The rationale for using tribe in place of deme instead of the commonly (mis)used term "a culture" is that the latter is being reserved for exclusive reference to an entity in the metaorganic information hierarchy and it is the common casual use of the same term for biological entities, their behaviors, and the information systems they share that contributes to the terminological confusion all this is intended to circumvent. The term "society," another commonly used alternative, is, as just noted, being reserved for a different kind of social entity in the economic hierarchy. More importantly, *when stripped of its organizational and evolutionism connotations* as is being done here, tribe semantically most clearly invokes human beings first and foremost, who just happen to share a culture, while also invoking the largest size persistent group with a shared identity based on a mythologized real or fictive common descent (under *H. sapiens*) that also tends toward at least some degree of endogamy in practice. In other words, it is a genealogical entity. Moreover, such a correspondence between what is here called a tribe and a deme already seems to be accepted by others (e.g., Durham, 1991: 342ff., where it is called a society; Richerson & Boyd, 1999), making it superior to alternative terms that either lack the same sense of bounded specificity (e.g., folk), have rather dark connotations (e.g., race), or are too obscure (e.g., ethnos, ethnie – see Bromley & Kozlov, 1989; Smith, 1986, 2000a).

The rationale for using society in place of population is briefly this. First, the term commonly refers to a geographically restricted, *interacting* populace, but not necessarily a genealogically self-identifying group, with a fluctuating membership that can potentially and often enough actually include individuals that are part of multiple different tribes—modern industrial societies being an obvious case in point. Second, and more importantly, the abovementioned interactions are characteristically economic interactions related to somatic well-being—e.g., exchanges based on a division of labor, which may or may not involve aspects of economic specialization, and not "more-making."

The problem now becomes that the jump from individual organism to deme (i.e., tribe) and population (i.e., society), as in Eldredge's hierarchies, is simply too large. Whether or not such intermediate levels are necessary for a proper purely biological understanding of other species is irrelevant. Such intermediate levels *are* necessary for a reasonably coherent understanding of human social organization, not to mention its underlying social structure, simply because reproductively and ecologically oriented culturally regulated human aggregations and associations of various

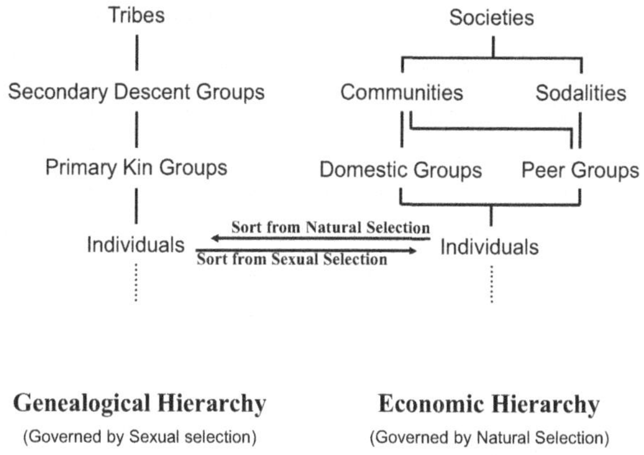

Fig. 4.1 The human organic hierarchies in social terms

intermediate sizes exist with sufficient regularity that they have often been considered products of cultural evolution, whatever that is variably said to be (e.g., Steward, 1955; see also Leonard & Jones, 1987; Johnson & Earle, 2000).

With respect to intermediate levels between the individual and both tribe and society, I propose to use two levels corresponding to Cooley's (1909) concepts of primary and (sub-societal) secondary social groups in both hierarchies (Fig. 4.1). The rationale for structuring the intermediate levels on this basis is that there are significant structural differences between the two types of groups and hence the social dynamics of the two levels. At the same time, they are general enough to leave ample room for the ad hoc use of sub- and supra-levels as needed, should the necessity arise for whatever reason in some specific hypothetical application.

Two points need to be clarified here. The first is that while secondary social groups tend to be larger than primary groups, as used here, entities at both levels are not defined by size but primarily by the nature of relationships within them (e.g., direct versus some degree of indirect interactions). Thus, it is conceivable that an instance of a primary group in one context can be larger than a secondary group in another context, particularly as regards groups in the genealogical hierarchy. The second is that it is theoretically possible for the individuals comprising a group in one hierarchy to be coterminous with the individuals comprising the same level of group in the other hierarchy, though in practice that is extremely unlikely to ever *precisely* be the case. In other words, both kinds of groups are highly variable in their specifics.

In any case, the two intermediate kinds of groups in the organic genealogical hierarchy, both characterized by real, adoptive, or fictive descent-defined relationships, are immediate kin groups such as localized lineage segments characterized by face-to-face relationships and extended descent groups such as non-localized lineages and clans, which are not. The rationale for this is that both are reproductively defined groups, and kin groups make more kin groups by segmentation. Aside from the difference between the primary and secondary group characteristics that distinguish them, the former roughly corresponds to an entity of a scale wherein kin selection can potentially operate at some level (e.g., Chagon & Bugos Jr, 1979) and the latter is sufficiently large that it cannot do so effectively. Family groups are not included in the genealogical hierarchy because even though they often function to serve reproductive purposes, they are most consistently characterized by a division of labor (i.e., economic interaction). Families may contribute to successfully making more individuals, who belong to such kin groups, but they do not make more families, and families are not even essential to successfully making more individuals because it is individuals—married or not—that ultimately make more individuals. As such, families properly belong in the organic economic hierarchy. So do what are called "houses" (e.g., see Lévi-Strauss, 1982). Though perhaps dominated by and named after a kin group, affiliation rather than descent is the ultimate determinant of membership, and their primary functions are arguably not so much reproductive as political and economic.

There are also two intermediate levels in the organic economic hierarchy. However, the complexity of human associative social behavior is such that doing even minimal justice to that complexity requires the construction of an ecological/economic hierarchy with two distinct types of groups at each intermediate level. The two intermediate levels are themselves again defined in primary and secondary sociological terms. However, within each level, there are groups that can be defined by residential association and others that cross-cut multiple such residential entities and are defined by common interest-based voluntary association.

Specifically, at the primary group level, we have both co-residential domestic groups (typically families) and localized small-scale peer/associative groups that cross-cut individual domestic groups. Both exist as constituent entities within the next level group of their respective type, with small-scale localized peer/associative groups also capable of being constituent groups within the next level residential type (the community). At the secondary group level, we have both communities in the residential branch and sodalities, age sets, trade groups, etc.,[3] that can cross-cut multiple communities in the associative branch. Both of these exist as constituent entities of the next level—the society. Unlike reproductive groups, all the groups in this hierarchy are involved in interactions aimed at enhancing the somatic well-being of their members in some manner and are typically characterized by some form of exchange, often involving some form of labor division.

Thus, shadowing Eldredge, the organic genealogical hierarchy for humans consists of gene, chromosome, individual, primary kin group, secondary kin group, tribe, and species (*H. sapiens*). Of these, only the individual through the tribe is of immediate concern for what follows; gene and chromosome are mentioned only for the sake of completeness, as is species at the other end. The organic economic hierarchy for humans consists of molecule, cell, individual, domestic group and localized peer/associative group, community and sodality, society, and regional

[3]Social classes are social categories, not social groups, and therefore abstractions and not entities. They engage in neither more-making nor regular interactions, nor do they carry information systems in their own right; only the individuals and groups so categorized do any of these. Thus, they have no place in any of these hierarchies.

biota. Here too, only the individual through the society is of immediate concern. Entities below the individual are mentioned only for the sake of completeness, as are the highest-order entities (regional biota).

Human Beings and the Entities of the Metaorganic Hierarchies

Metaorganic entities only exist for species with a repertoire of socially learned behaviors, and the larger that repertoire, the more elaborate the metaorganic entities are. For species whose behavior is completely programmed by their genetics, the determinants of behavior are completely situated in the organic hierarchies; there is arguably no significant element of purposeful decision-making affecting behavior beyond do/do not and therefore no true agency as the term is being used here. However, for species that do have a repertoire of socially learned behaviors, whatever its size, a new source of inputs to behavior exists alongside the programmed inputs from the organic hierarchies. That source is socially learned information, and it exists outside the organic hierarchies, as do now the behaviors that source influences. I leave it to others to decide how large a repertoire of such learned behaviors need be in order to be considered indicative of information systems large and complex enough to qualify for the label of culture. Suffice it for present purposes to say that humans carry ones that unquestionably so qualify since the term was coined to specifically label theirs.

While the entities of the organic hierarchies above the individual are structured by culture, they are essentially social entities consisting of two or more biological organisms. In contrast, what is commonly lumped under the heading of culture is metaorganic, or if one prefers Kroeber's (1917) term, it is "superorganic."[4] Cultural entities are, therefore, not organic and have no place in Eldredge's organic hierarchies, nor do the behaviors in whole or part based those cultural entities. Their proper place is in a new, second set of informational/evolutionary and behavioral/adaptive hierarchies respectively consisting of informational and informationally patterned interactive entities.

As with how to properly view the concept of species, the issue of whether cultures should be viewed primarily in behavioral (i.e., adaptive) or ideational (i.e., informational) terms has a long history of discussion within Anthropology (e.g., see Harris, 1968, 1979, 1999; Goodenough, 1981). To this has more recently been added the ethnographic version of the reductionist issue, in that some have come to question whether such things as cultures can even be properly said to exist at all (e.g.,

---

[4]I very much prefer metaorganic instead of Kroeber's *super*organic specifically to avoid the potential misconstrual of the latter term to imply a hierarchical relationship between the sets of hierarchies. All four hierarchies are structurally coequal. It is only each one's constituent entities that can be ordered hierarchically.

Abu-Lughod, 1991; but see also Brumann, 1999; Bashkow, 2004; Hanson, 2004; Varisco, 2018).

Issues of primacy in matters of culture change momentarily aside, virtually all anthropologists agree that there is both an ideational and behavioral aspect to what is *commonly called* culture (e.g., see Brumann, 1999: S3ff). Cultural information is transmitted in large part through the observation of learned behaviors much as a language is primarily learned by listening to others speak it. Ideas as to how to behave in a given situation normally generate normatively appropriate behaviors in the individuals who hold those ideas, and those behaviors generate ideas about how to act appropriately in the minds of others who observe the behaviors and their consequences in the context of a given situation (see Goodenough, 1981; Boyd & Richerson, 1985). But culture is no more transmitted by memorizing sets of context-specific behaviors, than language is learned by the memorization of context-specific stock phrases and sentences. Both are transmitted by observers abstracting out from observed behaviors the underlying deeply layered if-then decision "rules." For language-produced speech, those rules are called grammar; for learned behavior, they are called culture.

It is the information as to how to appropriately behave in a given situation that is transmitted, not the behavior itself, and the information system will evolve as competing ideas for how to behave in a given situation are differentially transmitted. So a culture can be said to be a socially transmitted information system that generates and regulates a learned, *flexibly adjustable* behavioral system that is the phenomenological expression of that information system, much as genes generate and regulate a biological organism *that is nevertheless capable of phenotypic plasticity* in its expression of those genes. However, even if we allow that it is information systems and their elements that must be what evolve by differential "more-making," learned behavior—the phenomenological expression of information systems and the primary medium by which those information systems are transmitted—must be fully accommodated in any attempt to define cultural entities.

The point here is that even though culture is generally adaptive and specific culturally patterned behaviors may be considered adaptations, a culture in and of itself is not an adaptive system (see also Durham, 1991: 361ff.). Rather, a cultural system should properly be seen as *a self-replicating reservoir* of reproductive and economic information. This reservoir is utilized differentially by entities in the organic hierarchies for their somatic well-being and successful biological reproduction. But it is the entities in the organic hierarchies, utilizing culturally conveyed behaviors, which phenomenologically adapt or become adapted by way of those behaviors. Were it otherwise, inasmuch as their economic adaptations were for all practical purposes identical, we would have to consider the historic Sioux and Cheyenne tribes, for example, to have shared the same culture.

Adaptations involve interactions with the environment or others, which by definition involve actions and thus behaviors. This has tended to force the logical position on many materialists that because human Culture (i.e., Culture with a capital

"C") is an adaptation[5] of our species and because adaptations must be phenomenologically (i.e., in the context of the current point, behaviorally[6]) expressed in order for matter-energy transfers to occur, culture must be defined phenomenologically (e.g., White, 1975; Harris, 1979). It further implicitly follows that individual cultural entities (whatever they are inconsistently said to be) must also represent adaptations.[7] Thus, they must also be defined behaviorally if one is to deal with the adaptive aspects of specific cultural systems.

Paradoxically, this is not true. That is, even though human Culture is an adaptation of the human species and even though specific culturally structured behavioral complexes may be adaptive (if not evolved adaptations), *a* culture in and of itself is not an adaptive system. Rather, it is *a self-replicating reservoir* of reproductive and economic information that gains and loses information over time—i.e., that evolves. This evolving reservoir of information is the basis for structured behavioral complexes that are employed differentially by various type and size culturally structured social entities in the organic hierarchies for their somatic well-being and reproductive success, and none of whom uses all the information contained in that specific informational pool (cf. Goodenough, 1981: 111ff.). But these behavioral complexes are not evolutionary entities; they are reproductively and/or economically adapted/adaptive, culturally structured behavioral systems utilized by various size culturally structured social (i.e., organic) groups. Accordingly, its adaptive aspects notwithstanding, a culture and its subsets *must* be defined first and foremost as self-replicating information systems, bounded (but not impermeably so) by some set of ideational parameters—i.e., a conceptual Bauplan[8] (see Rosenberg, 1994: 319ff; Bashkow, 2004: 453; and e.g., the concepts of "command pattern" in Miller, 1964 and "core tradition" in Boyd et al., 1997). Given the distinctively unique characteristics of cultural replication (see Boyd & Richerson, 1985, etc.), individual elements can, within certain constraints (e.g., the receptor culture's Bauplan, its environmental context, etc.), readily diffuse to other systems. This ultimately has repercussions for how one defines actual spatial boundaries (e.g., see Wolf, 1982), but that is a separate issue that will be addressed later.

Where does such an ideationally restricted definition of cultural entities leave the behavioral (i.e., interactive) and social (i.e., relational) aspects of cultural (sensu lato) systems? It simply requires them to be given different labels, as has already been done with culturally structured social entities. Cultures do not engage in behaviors—

[5]By virtue of having been selected for at the species level

[6]Morphological adaptations are rooted in the genealogical hierarchy and thus extraneous to this discussion.

[7]Ethnographers, faced with the messy and very visible reality of local-group variability, have tended to avoid broad adaptive generalizations with respect to cultures as entities (e.g., see Steward, 1938). Archaeologists, with no such inescapable corrective staring them in the face, have often fallen into that trap.

[8]Behavioral and social systems also have an "organic design," sometimes also called a *Bauplan* (e.g., Prentiss & Chatters, 2003), but the term will be used here exclusively to refer to a culture's organic design.

information systems cannot behave any more than genes have the capacity to be literally "selfish"; individuals and social groups of varying scales (individuals, families, local communities, kin groups, sodalities, societies, etc.) behave, as do multi-tribal associations. It is these social entities, variably utilizing cultural information and employing culturally generated behavioral systems, that adapt or are/become adapted and feed back into the mechanisms of evolutionary culture change. It is no more correct to speak of *the* American, Old Order Amish, BaMbuti, Natufian, Paiute, or Mayan economic adaptation than it is to speak of *the* adaptation of any given mammalian species, as noted previously (see Eldredge, 1985a, b: 159). It is this tendency on the part of materialists to see cultures as adaptive/economic entities that permits the use of the evolutionary socioeconomic and sociopolitical stages justifiably criticized by Evolutionary Archaeology and others (e.g., Leonard & Jones, 1987; Rosenberg & Rocek, 2019).

Having distinguished between human biological (i.e., organic) genealogical and economic social entities, the task now at hand is to do the same for cultural/culturally derived (i.e., metaorganic) informational and behavioral entities and then to systematically relate these four types of entities to each other (see Goodenough, 1999).

The Metaorganic Hierarchies: The Information Hierarchy and Cultural Entities

No doubt one reason for the widespread tendency among anthropologists to prefer skirting the issue of cultural entities is that the mechanics of cultural transmission makes cultural entities highly permeable to imported information by means of oblique transmission (e.g., see Boyd & Richerson, 1985; see also Bashkow, 2004; Hanson, 2004) and thus difficult to delineate—but not impossible to delineate (e.g., see Boyd et al., 1997; Mace et al., 2005). The real world abounds with evidence that cultural diffusion is not unconstrained, that cultural entities are in a very real sense bounded, in that acceptance of potential imports from outside the system is highly selective (see also Boyd et al., 1997; Henrich, 2004) and restricted to those novelties capable of being reconciled with and at least minimally capable of being integrated in some form or another into the existing system in its current (dynamic) state. Cultural systems are never fully integrated by virtue of being dynamic, but they do tend toward integration. As noted at the outset, the metaphor of an inchworm is appropriate, in that its tail end is always attempting to catch up to its front end but never truly does because its front end has moved forward again just as the back end has almost caught up to it.

Before addressing whether or not some kind of thing can be said to exist as an entity, we must first lay out what that thing is said to be. As noted above, cultural entities will here be defined as evolving information systems. It is widely held that the ideational aspects of culture—the information system—regulate the behavioral aspects (whether or not the ideational or behavioral aspects are considered to truly be part of a "culture" in any given theoretical paradigm). However, while it is the information system (which is here being labeled "culture") that must evolve, from a broader non-reductionist evolutionary perspective, a culture can no more be approached as simply an ideational system than a given species can simply be approached as a gene pool. Rather, strict definitions and labels aside, an evolutionary view of a culture must logically hold a culture to be an ideational system that *regulates* a behavioral system. If so, it follows that actual culture change *must* be defined as associated behavioral *and* ideational change. Without the former, there is no actual meaningful change, a neutral mutation so to speak. Without the latter, all one has is deviant idiosyncratic behavior, rather than cultural innovation. Deviant behavioral change may, as according to materialist theory, cause ideational change and thus precipitate culture change, but for culture change to actually occur, there must eventually be both behavioral and ideational change.

In those anthropological paradigms that acknowledge the existence of the things commonly called cultures, it is widely (though not universally) held that such entities are substantially integrated systems—complex *wholes*. This is not to say they are integrated in the static and fully integrated way in which they were implicitly presented by the functionalists of the early to mid-twentieth century (e.g., Radcliffe-Brown, 1952). Rather, it is to be seen more as a dynamic tendency toward integration essentially similar to what lies at the heart of Harris's (1979) view of "infrastructural determinism," wherein if one or more elements change, that change begets changes in other elements as the other aspects of the larger system change to reintegrate around the precipitating change. Full integration is never actually achieved simply because something in some part of the system is always changing, thus precipitating further integrative changes; and by the time those integrative changes fully play out, further precipitating changes have inevitably occurred, begetting yet further integrative changes.

It is also widely held within these same paradigms that the ideational rules, recipes, understandings, expectations, blueprints, templates, etc. (call them what you will), governing actual behavior are rooted in ideational beliefs and their derived values (i.e., what cultural materialists would call superstructure) that constitute the underlying justification and rationale for the employment of such behaviors. Thus, while behavioral change can precipitate ideational change, ideational components constrain behavioral change. Deviating from accepted behavioral norms incurs social penalties, which act to discourage such deviation, and the strongest penalties are applied for behaviors that violate the core beliefs and values that constitute the culture's underlying *(meta)organic design*—its Bauplan.[9] This makes superstructural congruity the primary cultural regulatory mechanism determining a potential innovation's acceptability and the primary cultural constraint on individualistic deviant behavior, importation of elements from outside the system, and cultural innovation. The systemic dynamics of culture change aside for now, a culture's Bauplan can, therefore, be said to be the central ideational principles of its

[9] Note the strength of the reaction to the attempted putsch of January 6, 2021, in the United States.

superstructural system, what are sometimes referred to as its eidos and ethos, not "matter-of-fact" usages or ideological justifications (see Wolf, 1982: 387ff.).[10] This is not to say that what is *commonly called* a culture is just an ideational system; as noted above, no such claim is made; it is to say that what are *here being labeled* a culture and its Bauplan are ideational in nature.

Its Bauplan is not the essence of a given cultural system (see also Boggs, 2004), nor is it a system's *external* boundary, any more than a building's steel girders constitute its essence or external boundaries; it is the system's structural core, just as a building's girders are its structural core. One cannot simply modify a building by adding to it in places where the existing girders do not offer structural support. Attempting to so modifying a building in a manner not supported by its girder structure will result in the building collapsing under the physical stresses produced by such additions. The same can be said of a culture. A culture's Bauplan is the information system's "girders" in the form of its conceptual core. A cultural entity can be said to remain *that* entity so long as the systemic integrity of its Bauplan is maintained. Loss of the information system's structural integrity, what Wallace (1972) referred to as a cultural "paradigmatic" failure, the more broadly cultural version of Kuhn's (1962) scientific paradigm shifts, results in disintegration of the system—its collapse—and the "death" of that cultural entity in the sense that it ceases to exist *as a system*. This opens the way to the emergence of what can be called a "new" or "descendent" information system (i.e., culture), one characterized by a different Bauplan. It matters not whether that new descendent information system continues to be commonly called by the same name as the one it replaces or not; it is a different system. Entities are not defined by their labels. Were it otherwise, there would be no point to this entire chapter inasmuch all the diverse things commonly called culture could legitimately be considered one and the same, making the need to properly sort them out nonexistent. Whether it is called by the same label as the system it descended from or not, the new information system is an entirely different entity, one characterized by a new Bauplan.

Leaving the issue of boundedness aside for the moment, cultural entities are ideational in nature and characterized by their capacity for being differentially transmitted—i.e., for evolving. They are *what* is transmitted, not how it is transmitted (cultural selection), who is doing the transmitting (biological entities), nor what is being done with the transmitted information (structuring behavior or organizing social relations). They are units of information at the most basic level, working up at higher levels to information systems held by individuals and information systems shared by members of varying type and size groups. They can be transmitted vertically, horizontally, or obliquely within such groups, as well as horizontally or obliquely to and from members of other groups (cf. Boyd & Richerson, 1985;

[10]Nor is a culture's *Bauplan* its religion as Miller (1964) proposed with respect to what he conceptualized as a culture's "command pattern." Such a general conceptualization of *Bauplan* is at once too rigid, too broad, and too vague, though some or even all of a culture's Bauplan can certainly be intertwined with the sacred.

Fig. 4.2 The human metaorganic hierarchies

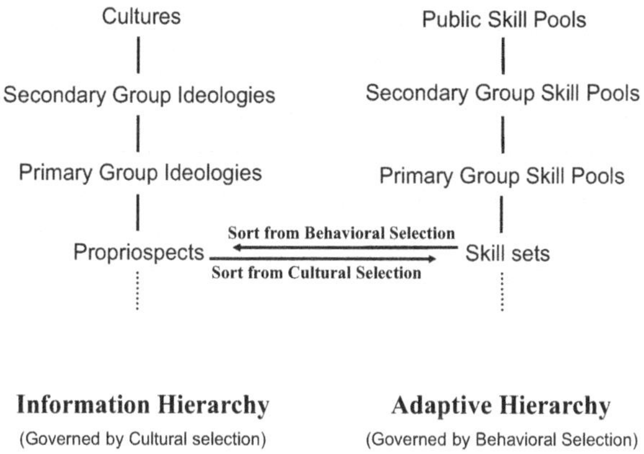

Shennan, 2002). But differential transmission is the selective process that drives the evolution of these entities; it is not the entities themselves. At each level above the most basic units of information, cultural entities are systems and thus constrained by their respective Baupläne. Obviously, it should be noted that just like any other aspect of culture, elements of the Bauplan can be understood variably by individuals, with self-interest often enough influencing such variability.

The most basic cultural (i.e., informational) entity will here be termed the "concept" (Fig. 4.2). The term "meme," coined by Dawkins (1989), has sometimes been used for such basic units (e.g., Durham, 1991), but it arguably belongs at the next level. As noted by Boyd and Richerson (1985: 37; see also Shennan 2002: 46ff.), the precise nature of such elementary units of cultural information remains poorly understood, so a formal definition is best avoided for now. However, as a tentative working definition, we would probably not go too far wrong if we consider concepts to be the information that can be encapsulated in the meanings of what are called linguistic terms—single words being too restrictive[11] —or their physical expression as gestures. More useful concepts are invoked more frequently than less useful ones and are thus transmitted more readily than less useful ones.

At the first level of the information hierarchy above the concept, we arguably have the proper place for the meme, constituting a conceptual relationship—i.e., a set of concepts related in some way (e.g., send-letters-get-luck, in the case of the chain letter example used by Goodenough & Dawkins, 1994 to illustrate a "selfish meme"). This usage of meme seems more in keeping with Dawkins' apparent structural conception of the term—the relational aspect being a crucial element in a meme's ability to propagate itself. The danger with actually using the term "meme" for this level of informational entity is that it is commonly applied in everyday English to such a wide variety of things that the only thing they have in common is their high level of transmissibility rather than the structure of their informational

[11] For example, "*arm* chair" is a term that cannot be expressed as either the word "arm" or "chair" and for which no single word has been coined in English.

content. Since neither concepts nor memes will substantially enter into any of the discussions to follow, the term "meme" for this level of entity can be left in place for now.

Above the meme, we have the first relevant systemic entities: the individual's "propriospect" (Goodenough, 1981: 98), defined as an individual's "private, subjective view of the world and its contents." This propriospect is not simply *the sum total* of the concepts and memes that are known to a given individual. Rather, the propriospect is the sum total of *integrated* information accepted as *valid* information by that individual. Validity, in turn, is determined by the Bauplan of the individual's propriospect, which is an individualized variant of all applicable (to that individual) higher-order Baupläne (see also Handwerker, 2002). For example, if that propriospect's Bauplan includes concepts of supernatural forces, then witches, ghosts, miracles, and/or magic are potential realities within that information system to be used or addressed as needed. If it does not, then they are not valid information (to be used as needed). This is because they are not real to the individual, even though that individual may be aware of *purported* examples of witches, ghosts, miracles, or magically induced events and aware of others who do consider them real.

The next three levels of the cultural hierarchy involve ever more inclusive collective information systems that generally propagate by fissioning in the context of group social dynamics and which no single individual's propriospect encompasses completely. As noted in the previous section, there are three different organizing principles at work in social group formation. They are kinship, residence, and (broadly speaking) common interest-based association. The first characterizes social entities in the organic genealogical hierarchy, and the latter two characterize entities in the organic economic hierarchy. All three of these group types can potentially carry information systems, but at any given level, not all three types will necessarily carry *persistent* information systems, though they can. It should also be noted that the number of distinct information systems that can exist at each level within each of these three tracks varies with social complexity (i.e., number of groups, each carrying its own) and that an individual can belong to multiple groups at each level, accepting as legitimate the basic elements of their respective information systems, even if there are conceptual conflicts inherent in the acceptance of any number of them.

Such group-level information systems exist as additive overlays on the individual's propriospect, to the degree that the individual identifies as a member of that group. Transmission of the shared system is not the same as the individual adopting one or more elements of that system or even pieces of unrelated information from other individual group members to incorporate into their own propriospect, even though the mechanics of transmission can be the same. The former deals with what it is to be a member of that group (i.e., "This is what *we* believe/value" and "this is how *we* do things") and the latter with what it is to be an individual person ("this is what *I*...; this is how *I*...."), who may or may not incidentally be a member of some number of groups. Such group-level (i.e., shared) information systems may complement, conflict, or be neutral with specific elements of the individual's propriospect, or with other group-level information systems in cases of multiple associations on the part of a given individual. Conflicts between elements of two or more systems accepted by an individual require context-specific compartmentalization, with the question of which of the conflicting elements takes precedence being determined by whether the individual identifies him/herself primarily as an individual or a member of some specific group *at that particular moment* (see also Goodenough, 1981: 99). Group-level cultural entities often appear to be only abstractions because of the potential multiplicity of sub-entities and the variability of their Baupläne (which share an increasingly more limited number of elements the higher in the hierarchy one looks), but that is because intermediate entities between the individual and the highest-order entities are ignored (see above).

In any case, at the third or primary group level, there are first primary kin group ideologies, carried by primary kin groups in the organic genealogical hierarchy. Such information systems are a property of primary groups consisting of real, fictive, or categorical siblings and their offspring and are generally transmitted vertically from senior members (e.g., parents, parents' siblings) to junior members (e.g., children, nephews/nieces) and, to a lesser extent horizontally, between junior members. Such systems propagate by fissioning, as lineage segments break off. The coherence, stability over time, and very existence of any such primary kin group information systems vary with the type of descent system and residence norms employed by the groups in question—i.e., the relative exposure to one or another parent's primary kin group by junior members. Thus, at one extreme, such systems will presumably be most temporally persistent, and vertical transmission will presumably occur with the greatest fidelity in the case of unilineal descent groups where the family resides with the descent group with which the child both is identified and identifies—and into which the resident alien parent will tend to become at least somewhat acculturated. At the other extreme, it will likely be more temporally ephemeral and vertically transmitted with less fidelity in cases of bilateral descent coupled with neolocal residence.

In bilateral-neolocal social systems, such as are common in western industrial societies, there is in a very real sense no *persistent* information system to speak of being transmitted in the kinship track at all. Instead, an information system is cobbled together in the residential track after marriage from the often different propriospects of the marriage partners, and it lasts only for the duration of the marriage. One expression of this process is the commonly recounted anecdotes within adult peer groups about how in the early years of a marriage there were often heated arguments about the proper way to celebrate important holidays (e.g., Christmas in the west) or some other important events ("But we did it *this* way in my family, not your way"). Such conflicting information systems necessitate compromises that produce some hybrid of the two "ways," which then becomes *the* "way" for the children born and enculturated into that family until they too compromise it away in adulthood with their own marriages. This extreme cultural "crossing-over" in the formation of such primary kin group ideologies has repercussions for the evolution of cultures with such marriage systems, but that is a tangential issue (see Boyd & Richerson, 1985).

It is worth emphasizing that a person learns information from both parents by vertical transmission. Individual elements are incorporated into that part of the individual's propriospect pertaining to him/herself as an individual, while other elements associated with one parent's descent group (or both, in cases of duolineal descent) are incorporated into that part pertaining to membership in the group(s) *and that is a different matter entirely*, as is any information incorporated into the propriospect obliquely or horizontally from others outside the residential group.

The information systems associated with the two primary groups from the organic economic hierarchy—peer and domestic—are equally problematic, in that more often than not they are also temporally ephemeral and do not persist as distinct components within higher-order evolving information systems. More often than not, primary peer groups (including noncyclical localized sage sets) exist for relatively short spans of time, with any emergent information system that coalesces during that interval "dying" at the dissolution of the group. Localized cyclical age sets, of sufficiently short interval to allow for identification with and interaction between living members of a repeated set, would be an obvious potential exception. Small hunter-gatherer bands, where the adult male or female peer group corresponds to the entire gender-specific population of the community, are another important exception. This is because the adult gender peer group persists (along with the information system it carries) as new, young adults are brought into it at maturity. Moreover, hierarchically, at the next higher level, the community-level group's information system can be said to encompass the two persistent gender peer group information systems.

Domestic group information systems are more variable. That is, at one extreme, domestic groups will tend to carry the dominant kin group's ideology in unilineal-paralocal social systems, while at the other extreme, they will be, as noted above, as transient as those of most peer groups in the case of bilateral-neolocal social systems. In neither the case of peer nor the latter type of domestic groups can it be said that such type groups are *consistently* characterized by a shared world view *that is propagated* in some fashion because peer groups typically do not make more peer groups and domestic groups typically do not make more domestic groups. Such ephemeral emergent information systems do exist as constituent entities within a given culture, and the emergent information systems of such ephemeral groups *do* feed into the developing propriospect of the individuals who are members of such transient groups and thus into the evolution of higher-order cultural systems. However, their differential propagation as distinct entities is not always (or perhaps even often) an element in the evolution of higher-order informational entities.

The fourth level of the information hierarchy is that of secondary group information systems, with those of sodalities,[12] societal-spanning age sets,[13] etc., existing as entities in the voluntary track; those of communities existing as entities in the residential track; and those of extended kin groups (e.g., extended lineages, clans) in the kinship track. These exist as a further overlay on the lower-level information systems that individuals accept as valid. The "corporate cultures" often mentioned in the context of industrial/postindustrial economies are an example of such voluntary-track information systems, so are the information systems held by voluntary associations more traditionally studied by anthropologists. Again, propagation is by fissioning of the carrying group when it happens.

Such secondary group information systems exist as a further level of overlay on lower-level systems (e.g., "corporate culture" as an overlay on a coworker peer group/department information system, which in turn is an overlay on individual propriospect). Again also, an individual can subscribe to multiple information systems at this level and compartmentalize horizontally and vertically as necessary when information systems conflict.

All three tracks converge again at the fifth level, that of an information system that is a culture, representing the highest level of shared information system—one carried by a tribal group or a society if that society is constituted by a single tribal group or, more commonly, dominated by a specific such group. This is the highest-level possible overlay on lower-level systems, and its Bauplan is what carries down in whole or part through all the lower-level systems in variable form. Thus, a carrier of some culture can be said to subscribe to this highest-level information system and also to the various lower-level systems that are the property of lower-level groups to which that individual belongs and ultimately down to that individual's propriospect. It should be noted that in cases of imposition, lower-level information systems may lead their carriers to both not subscribe to and contest the higher-level information system, but they are still governed by it so long as they participate in the groups that carry the higher-level system, whether that participation is voluntary or not.

The sixth and highest level is that of Culture (culture with a capital C), constituting the full human culture pool (see Goodenough, 1981). As with the corresponding organic entities (humanity and regional biota), it will not enter further into this discussion and is being mentioned here, along with concept and meme, simply in order to fully flesh out this hierarchy of entities.

One important point needs to be repeated. A person is a member of a society and tribal group but subscribes to cultural systems. Moreover, while a tribal group

---

[12] Societal-level social groups whose membership is defined, not by (real or fictive) common descent or by common residence but by voluntary or involuntary association in order to engage in activities with a common ultimate purpose, that can cross-cut multiple instances of societal-level social groups organized around the first two defining principles

[13] The ephemerality (as systems) of the information systems carried by noncyclical age sets of any scale is evident in the transience of the various "youth cultures" (e.g., mod, rocker, hippy, punk, skater, goth, etc.) that began to appear in industrial societies starting in the second half of the twentieth century.

arguably carries a single culture, a society does not necessarily do so (e.g., BaMbuti-Lese) unless constituted by a single tribal group or dominated by the culture of the dominant tribal group.

An important point also needs to be clarified here. Information systems are not homogeneous at any group level. Particularly in complex societies, as noted above for societies, the culture of a tribal group can be dominated by the information system(s) carried by one or more of its constituent secondary groups, which in turn may be dominated by the information system(s) carried by one or more of the secondary group's constituent primary groups.

Another important point also needs to be clarified here. It is that at this level also, a person can potentially adhere to multiple systems as long as they are not mutually exclusive in terms of their systemic structure, with the individual again compartmentalizing horizontally and vertically as necessary (see Goodenough, 1981, 1999). In other words, a person can be an adherent to multiple cultures (i.e., tribal group-level information systems), and not all groups living within a society (particularly large multi-tribal state societies) are necessarily adherents to information systems that can properly be said to be subsets of some higher-level information system. They may, in some cases of what are commonly called *sub*-cultures in modern multi-tribal industrial societies, adhere to information systems that have no direct overlay representing the higher-level information system. Thus, in reality, such individuals simultaneously adhere to multiple independent *different* cultures that may only share a common language (if that) and share little else with the dominant tribe's culture besides units of information that are nevertheless organized quite differently—sometimes to the degree of outright conflict in the case of what are commonly called counter-cultures.

For example, adherents to all the various hyphenated American subcultures (e.g., Italian-American, African-American) are arguably in reality adherents to two, typically non-mutually exclusive, cultures. The first is typically a peripatric breakaway of some more spatially distant culture that has evolved in an environment dominated by another culture to be generally adapted to some degree to that dominant culture. The second is the dominant culture itself. Adherents to such "subcultures" are actually simultaneous members of two cultures and shift back and forth and compartmentalize as necessary. The fact that such adherents to multiple cultures may interact regularly within the framework of a single societal system does not mean they are necessarily adherents to a single overarching information system, nor does the fact that they may speak the same language.

Lastly, two points need to be stressed. The first is that just as individual propriospects and individual understandings of group-level information systems are variable (and all variability is subject to selective forces), there can be variability between lower-level systemic "integrations" of the various higher-level elements that exist as overlays on them. Such systems are, after all, entities—i.e., individuals—and variability in their respective (variably shared) "understandings" is to be expected (see also Handwerker, 2002). There can even be variability between and within systemic levels concerning the understanding of higher-order Bauplan elements—as between higher-level American and lower-level KKK[14] or NAACP[15] understandings as regards who qualifies for inclusion under the Bauplan principle of "equality" and even the details of what "equality" specifically entails, as manifest in the continuing tensions concerning affirmative action social policies a half century after they began. Higher-order cultural entities, such as a culture, appear to be abstractions only because of the multiplicity of sub-entities that potentially fall under them and the variability of such sub-entities. The greater the vertical gap between any two such hierarchical entities, the more the higher-order one appears to be to an abstraction. This is most pronounced when the gap is between the individual propriospects and a culture, but that is only because intermediate entities between the individual propriospect and the culture are being ignored.

The second is that lower-level systems are not invariably strictly speaking subsets of higher-level systems, in that not all the information contained in any lower-level systems is necessarily incorporated into higher-level systems. This is a multilevel version of what Goodenough (1981: 111) distinguished as the difference between a society's "public culture" and "culture pool." Information from some lower-level systems may be known to other adherents of the higher-level system that nevertheless adhere to different lower-level systems, but that is different than accepted as a legitimate part of the higher-level system (see above). Higher-level systems are thus not of necessity true sums of all the lower-level entities, particularly as regards entities in the economic hierarchy.

The Metaorganic Hierarchies: The Adaptive Hierarchy and Behavioral Entities

Behavioral metaorganic entities are culturally derived, learned, and ultimately for the most part economic and reproductive related behaviors. They too have a Bauplan so to speak, revolving around the specific fundamental purpose of the behavioral system (see Prentiss & Chatters, 2003). However, as used going forward, Bauplan will refer to the cultural and not the behavioral Bauplan, which should probably be given a separate term. They are the phenomenological expression of ideational systems, the actions, not the actors, nor the information the actions are based on. The adaptive hierarchy of which they are part differs from the information hierarchy in the fundamental respect that the entities which exist at its various levels above the most basic consist of goal-oriented behavioral systems designed to interact with other such systems and/or the environment in order to achieve some agentially driven purpose. They are largely based on information derived from the various

[14]The Ku Klux Klan. This was a secondary group that was an important political power at the local level in many parts of the United States during the first two-thirds of the twentieth century.

[15]The National Association for the Advancement of Colored People. This was an important secondary group in the twentieth century African-American civil rights movement.

levels of the information hierarchy, but they are not themselves information or information systems. They are culturally structured behaviors whose components are designed to accomplish specific ends, which for the most part ultimately bear on economic and reproductive success. The degree to which such ends are successfully accomplished and the degree to which any such success or failure is judged by others to be a product of the underlying information's role in determining the outcome (as opposed to other factors, such as chance, the practicing individual's competence level at the behavior, etc.) affect the reproductive success of the information (in the evolving information hierarchy) underlying the behavior, just as it biologically affects the practicing individual's reproductive and/or economic success in the organic hierarchies.

The basic unit in the adaptive hierarchy will here be called the physical action, representing some specific basic form of learned or natural physical movement (Fig. 4.2) capable of producing an environmental effect.

Above that is the skill, constituting some learned set of related actions mastered through actual practice and designed to produce a specific material effect. A skill is not simply the knowledge of how to do something; it is the reasonably *successful doing* of it with some minimal degree of competence (see Goodenough, 1999: 98). Obviously, knowing how to do something (information) is essential to being able to do it, but knowing how to do something is nevertheless not the same as actually doing it competently. It should also be noted that not all skills are necessarily directly related to economic or reproductive pursuits; they can also relate to cultural reproduction (see Cronk, 1999: 92).

The third level is the individual's skill set. This constitutes the sum total of learned skills possessed by an individual and through which a culture-bearing individual can interact as needed with their physical and social environment to purposefully achieve an effect. It varies with individual culture bearers, being a product of the various information systems they subscribe to, their competence, and the demands of the social and physical environment for actual actions. Thus, as in the organic hierarchies, the individual (in the form of his/her cultural and behavioral proxies as propriospect and skill complex) is also the nexus of the two metaorganic hierarchies. And it is this straddling of all four hierarchies by an individual culture bearer's organic self and metaorganic avatars that fosters the confusion regarding entities.

Above the individual skill set in the adaptive hierarchy is the primary group skill pool. It is the set of redundant and complementary skill sets carried by the individual members of primary groups in the economic (domestic and peer groups) and reproductive (primary kin groups) hierarchies. This is the lowest level of behavioral complex that can involve aspects of exchange. Such exchanges can be reproductively oriented, as in the exchange of individuals in marriage, or economically oriented, or some intertwining of the two.

Above the primary group skill pool is the secondary group skill complex, constituting the multiple immediate kin, domestic, and peer group skill pools carried by secondary group entities in the reproductive (clan) and economic (communities and sodality) organic hierarchies. It is the exchange aspect of these secondary group-level (e.g., community) behavioral systems that allows for skill specialization at both the individual and primary (notably domestic) group levels. It is also cooperative participation in this behavioral system that allows for the true long-term economic and reproductive success of its constituent primary group organic entities.

It is typically at this level of entity—specifically an organic community and its metaorganic *economic* skill pool—that we can realistically talk about a culture-bearing group's ecological niche. For example, the Paiute (as a tribal group) did not occupy a specific ecological niche in any meaningful sense of that term; individual communities or groups of localized communities did (e.g., Owens Valley Paiute). The same applies to neighboring groups in other valleys (see Steward, 1938). However, while the economies of neighboring individual communities may have been somewhat different, an individual transferring between any two would have had minimal difficulty adjusting socially or economically to the transfer. This is because typically the members of adjacent communities tended to share a similar if not common culture (i.e., share a single overarching information system). Thus, all that is necessary for the appropriate skill(s) to be added to the individual's behavioral repertoire (skill set) is practice in the hitherto unutilized behavior that was already known to them for being part of the shared information system. This flexibility of potential behavior is an expression of what we can call that specific *culture's* extended behavioral phenotype, to appropriate that evolutionary term and apply it to the metaorganic hierarchies. A true outsider would first have to be acculturated (i.e., learn and accept the relevant information as valid), before being motivated to hone the requisite skill(s) through practice.

At the next level above the secondary group skill pool, there is the public skill pool, constituting the behavioral repertoire available to a society or tribe, by virtue of its constituent secondary group, primary group, and individual skill complexes. It is a regional network of secondary group skill pools tied together for purposes of economic and/or reproductive ends. It encompasses neighboring communities engaged in complementary or competing behaviors involving energy or matter exchanges that aim to enhance the participants' somatic well-being beyond that minimal level required for simple survival, as well as their reproductive success. It is not necessarily restricted to encompassing just communities that share a common culture, though it can be, as in the case of hunter-gatherer macro bands. It is this level of behavioral interaction we mean when referring to regular BaMbuti exchanges of meat and honey for produced plant foods with neighboring Lese, or producing communities interacting with administrative centers in state systems. While we may not be used to thinking of such disparate social entities such as BaMbuti and neighboring Lese communities as being members of a single society, such a view is easily supported by the high degree of interaction and interdependence documented by ethnographers (e.g., Bailey, 1991). All that is missing for us to "recognize" it as a society, in the typically used sense of the term, is a unifying *formal* political structure. Such informal societies are nevertheless societies in any economic sense of that word, and the organic economic hierarchy consists of economic entities, formally organized or not, not cultural, behavioral, or genealogical ones. The point again is that it is not the informational or organic genealogical entities (i.e., *the*

BaMbuti, for example) who engage in any specific such economic exchanges, but specific communities of organic entities and not necessarily all of them at that, or in necessarily the same manner (e.g., see Bailey, 1991).

Lastly, there is the human behavioral pool. This constitutes the full range of human behaviors stemming from human Culture and corresponds to the highest level of the other three hierarchies. As with the case of the other three hierarchies, this highest level (along with action and skill) will not enter further into this discussion, and they too are again being mentioned here just in order to fully flesh out this hierarchy of entities.

Archaeology and the Hierarchies

Though not needed for present purposes, there are two issues that should be addressed in order to integrate archaeological remains into this conceptual framework of hierarchies, if for no other reason than archaeological remains are the actual record of cultural evolution. The first is where does material culture in general and more specifically where do artifacts fit into this framework of entities. The second is how the different kinds of culturally patterned entities can be recognized on the basis of artifactual remains.

In the matter of the former, the long and widely repeated dictum that "pots are not people" makes the point that there is a distinction between artifacts and the organic entities that produced them. To that, I would add that neither are artifacts ideas, nor are they behavior. Rather, material culture and the individual artifacts that constitute it are the extended phenotypical manifestations of cultural phenomena; they are the second-order metaorganic manifestations of ideas, behaviors, and the metaorganically organized organic entities that made them. Thus, the proper way to view material remains is as a second-order "shadow" set of four identical hierarchies consisting of the artifactual manifestations of the first-order hierarchies. This in turn requires some way to relate manifestations of the second-order material entities to the first-order cultural and culturally structured entities that produced them.

Admittedly, a construct of four hierarchies plus their four second-order archaeological shadows is a complex way to view culture, but I offer the same counter as I did to Gould's (2002a) criticism of Eldredge's biological hierarchies (see Sect. 2.5). I would argue it is not *unnecessarily* complex; in fact, it abstractly describes an even more complex reality. I would further argue that it provides a set of tools that are needed to facilitate discussion and simplify the path to understanding.

For example, the disagreement between Boone and Smith (1998) and Lyman and O'Brien (1998) can be clearly understood with the aid of this heuristic to be much more a case of "apples and oranges" than first appeared when discussing whether natural selection can properly be said to operate on the evolution of artifacts. That is, Lyman and O'Brien (1998) were very likely referring to natural selection's effect on cultural selection [on the basis of the shadow hierarchy's perceived contribution to the user(s)'s fitness] and then cultural selection's effect on behavioral choice, but they had previously presented it as natural selection affecting behavioral behavioral choice and taking place in the *shadow hierarchies*. On the other hand, Boone and Smith (1998) were referring to the same processes [observable behaviors' contribution to the actor(s)'s fitness] but taking place in the *primary hierarchies*. In other words, they were both in a very real sense correct in that they were both referring to something real. They were just talking about different parts of the proverbial elephant's anatomy, with Lyman and O'Brien more importantly also initially leaving out a crucial step in their process (the relationship of the shadow to the primary hierarchies) because I would argue they did not have a fully labeled description of that anatomy readily at their disposal.

In any case, with respect to archaeological signatures, artifacts are made by individuals or groups as expressions of the corresponding level metaorganic entities or the material residues of social organization. While artifactual entities lower than the artifact (e.g., attribute/trait) do exist, for present purposes, they have no meaningful counterpart in the first-order hierarchies, being meaningful only in terms of describing and/or defining artifacts themselves, including their possible functions and/or evolution. Thus, in the second-order (artifactual) shadow hierarchies, they can once again be removed from consideration with respect to what follows. However, it should be noted that despite my emphasis on the systemic nature of the entities being discussed, one could use this construct to argue that the reductionism of Evolutionary Archaeology is actually warranted as regards its analytical focus. This is because it is dealing with entities below the individual's artifactual proxy (i.e., traits, produced by skills) and thus minimally systemic, if even that.

Archaeologists have for the most part always intuitively grasped that there was some form of relationship between decorative styles and cultural entities of one kind or another. However, they generally failed to systematically distinguish between "culture" as an information system, "culture" as a genealogically or economically oriented social group, and "culture" as a system of behavioral patterns. The application of the proposed hierarchies to the archaeological record obviously necessitates addressing the archaeological signatures of the various types of entities being proposed. It should be stressed that what follows refers to the kinds of entities, as defined by the hierarchy in which they occur, not the level of a given group within a given hierarchy. That is, while distinguishing between individuals and groups in the archaeological record is generally important, the importance of determining the level of group within a given hierarchy is not always equally important, being problem dependent and thus variable.

A detailed discussion of this methodological issue is both too complex and too tangential to attempt at this time. Suffice it for now to offer the following generalizations. The economically functional aspects of artifactual residues can generally be considered (extended phenotypical) expressions of various level entities in the organic economic hierarchy, be they individuals or groups. The symbolically (i.e., non-economically) functional aspects of artifactual residues can be considered to constitute "boundary markers" and thus expressions of various level entities in the organic reproductive hierarchy (e.g., see Boyd & Richerson, 1987), though when a society is constituted or dominated by a single tribe, such boundary markers can define such societies as well.

However, the patterned *associations* inherent in artifactual residues of any kind are generally expressions of metaorganic behavioral systems of some level engaged in by organic entities, but not the organic entities themselves. And lastly, symbolic *systems*, as in the form of decorative complexes, are the visible expressions of integrated metaorganic information systems, of which the aesthetic principles from which such symbolic systems are derived are part. They are another set of boundary markers that allow us to delineate the spatiotemporal bounds of the information system of which they are part, and the analyses of artifact functions and associations may perhaps permit the delineation of some of the specifics of that information system of which they are part.

Chapter 5
That Complex Whole and its Evolutionary Dynamics

"Evolution is the fate of transmissible information in an economic context."—Niles Eldredge1

The Hierarchies and their Modes of Selection

Both Eldredge's view of hierarchical dynamics and Walsh's situated Darwinism mean that the four hierarchies outlined in the previous chapter do not each exist in an evolutionary vacuum. They are all paired with each of the others in a series of complex dynamics that tie the metaorganic domain to the organic (social) domain. Having taxicly delineated the various proposed kinds of entities in question, we can now begin to examine these evolutionary dynamics.

To begin with, these four hierarchies are structurally coequal. Selection takes place within each hierarchy on the entities that constitute that hierarchy, and each hierarchy is characterized by a distinctly different mode of selection. In the organic hierarchies, as noted previously, those modes are natural selection operating in the economic hierarchy and sexual selection operating in the genealogical hierarchy. Natural selection refers to the natural and social environment favoring some organic entities in the economic hierarchy over others of the same or other kinds on the basis of phenotypic traits, including social ones, that function to enhance somatic well-being. Sexual selection refers to the natural and social environments favoring some organic entities in the genealogical hierarchy over others of the same kind on the basis of phenotypic traits, again including social ones, which function to enhance organic reproductive success.

In the metaorganic hierarchies, the modes of selection are cultural selection operating in the information hierarchy and what we can call behavioral selection in the adaptive hierarchy. This last term is borrowed from Behavioral Psychology, where behavioral selection is used interchangeably with operant selection to refer to behavioral choice (e.g., see Skinner, 1981; Hull et al., 2001), which is precisely what the mode of selection operating in the adaptive hierarchy involves—one potential behavior being chosen over others in a given context by a given entity. Thus, both involve choice. Cultural selection refers to acceptance, or lack thereof, by an organic

M. Rosenberg, *The Dynamics of Cultural Evolution*, Studies in Human Ecology and Adaptation 12, https://doi.org/10.1007/978-3-031-04863-0_5

entity of interpreted information for incorporation into the information system it carries. It refers to the choice of what to believe (incorporate into the information system) versus what to reject, and such choices are by no means always objectively correct since they are influenced by multiple extraneous factors, such as confirmation bias. Behavioral selection refers to considered decisions concerning the choice of actions and/or skills to be employed in specific environmental contexts in order to achieve a desired end. In other words, they are purposefully intent driven and thus agential, whether or not one or more of the behavior's objective functions are what the behavior is actually intended to accomplish. Here too also, the choices made are by no means always objectively correct, much less optimally so.

Cultural selection and behavioral selection both include significant Lamarckian elements in their operation in addition to Darwinian ones. However, as noted above, this does not by any stretch of the imagination imply that these selective processes are necessarily perfect or even consistently close to correct in some objective sense. This is because both are filtered through existing cultural understandings that can potentially distort or obscure objectively true realities, leaving natural and sexual selection to weed out the objectively poor decision makers.

Selection, Sorts, and Hierarchical Dynamics

As per Eldredge's solution for finding a path through the biological version of the entity problem, the sort from the human economic hierarchy, produced by natural selection operating on economic entities, flows to the human genealogical hierarchy at the level of the individual, supplying favored economic variants to that hierarchy for sexual selection to act on, while the sort from the genealogical hierarchy, produced by sexual selection favoring reproductive variants, simultaneously flows to the economic hierarchy for natural selection to act on what it supplies. Given the clear parallels (but not identity) in the informational-phenomenological (i.e., replicator-interactor) structure of the organic hierarchies and the metaorganic ones, processes at work in the organic domain, but not logically restricted to it, can logically be expected to also be at work in the metaorganic domain. Thus, the same dynamic, with the sort from one hierarchy simultaneously flowing to another hierarchy for selection in the destination hierarchy on the basis of a different set of traits, can also be said to operate in the two metaorganic hierarchies (see Rosenberg, 1994). However, the actual hierarchical dynamics now are not quite as simple as either this last statement or my initial attempt to address those dynamics (Rosenberg, 1994) would appear to suggest. This is because, since we are dealing with both organisms and culture (sensu lato), there is more than one destination hierarchy for any given sort.

Thinking of these dynamics in the metaorganic hierarchies just in terms of the usual ideational-behavioral opposition with one feeding into the other and vice versa, the way in which the structural dynamics of culture are usually viewed superficially suggests a neat parallel to the dynamics of the organic hierarchies. In one direction, the sort from cultural selection in the information hierarchy flows to the adaptive hierarchy at the level of the individual culture bearer and provides the range of *acceptable* behaviors available to the individual to choose (or deviate) from. In the other direction, the (perceived) degree of relative success ascribed to a particular behavioral choice at achieving the intended goal of that behavior—i.e., the sort from the adaptive hierarchy—flows back to the information hierarchy where it serves as the basis for the potential propagation of the information underlying that behavior.

While such a dynamic does cover the mechanics of cultural transmission (e.g., Boyd & Richerson, 1985), the problem lurking here is: relative success at what? Such a simplistic view of evolutionary cultural dynamics includes no yardstick against which to measure the relative success of behaviors and thus the frame of reference cultural selection requires in order for it to operate. That yardstick requires reference to the organic hierarchies, which is where all forms of evolutionarily *meaningful*[1] biological success are measured. After all, Culture is an adaptation of our species. That is, the two metaorganic hierarchies do not exist in isolation from the organic hierarchies, the human entities that are the carriers of information systems and typically behave in approximate accordance with them, nor do the organic hierarchies exist in isolation from the metaorganic hierarchies, the socially learned components of human adaptations; humans and culture are, after all, coevolved. Cultural information and culturally structured behavior can no more be properly understood in evolutionary terms without reference to the organic entities they exist to benefit either reproductively or somatically (and without whom they cease to exist) than can the human entities without reference to cultural information and culturally patterned behavior.

The point is that the sort from any one hierarchy does not just flow to a single other hierarchy—organic to organic and metaorganic to metaorganic. Elements of the sort from any one hierarchy in fact flow to *each of the other three* hierarchies at the level of the individual or the individual's metaorganic proxy (propriospect and skill complex), making the individual not just the nexus of the organic hierarchies, but of all four hierarchies. This arriving sort provides the raw material for the selective forces respectively operating within all three of the other hierarchies. Thus, there are actually 12 different individual pairings of these four different hierarchies (Figs. 5.1) based on the hierarchy in which the sort originates and the one to which it flows and in which it terminates. In order to deal with these dynamics, designations for these sorts are necessary, and the designations that I will use here for these various pairings will be an abbreviation based first on the mode of selection that produced the sort followed by the mode of selection it is flowing toward to be further acted on.

---

[1]Thus, excluding fads, fashions, and other forms of drift, which are only successful at replicating within the metaorganic hierarchies

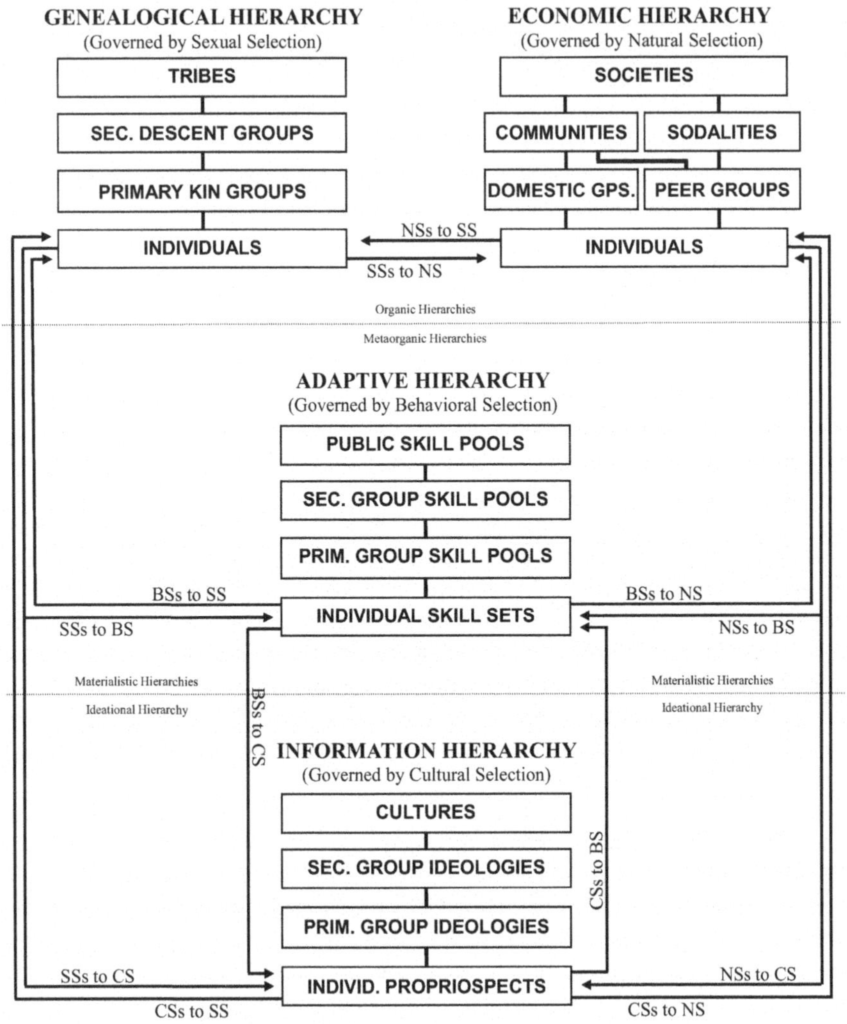

Fig. 5.1 All four hierarchies and their dynamics

Natural Selection and its Sorts

Natural selection in the economic hierarchy provides the following pairings to the other hierarchies (Fig. 5.2a):

NSs→SS: the *n*atural *s*election *s*ort from the economic hierarchy being made available to *s*exual *s*election in the genealogical hierarchy

NSs→BS: the *n*atural *s*election *s*ort from the economic hierarchy being made available to *b*ehavioral *s*election in the adaptive hierarchy

NSs→CS: the *n*atural *s*election *s*ort from the economic hierarchy being made available to *c*ultural *s*election in the information hierarchy

As noted by Eldredge (and also Hull, 1980), NSs → SS determines what is available in the way of biological individuals for sexual selection to act on in the organic genealogical hierarchy, based on the sort of organic entities that natural selection favors. On the other hand, NSs → BS is the sort produced by natural selection operating on biological entities with specific reference to behavioral systems' relative contributions (as opposed to physical attributes—see sexual selection below) to those components of fitness involving somatic well-being (i.e., longevity, economy of energy). This sort expresses itself in the adaptive hierarchy's decision-making processes as the *perceived* economic efficacy of alternative behaviors. Lastly, NSs → CS is the sort based on an individual's relative somatic well-being that flows to cultural selection as understandings about economic cause-effect in the natural world and on which cultural selection acts by way of differential transmission of economic and technological information (as distinct from behaviors).

Sexual Selection and its Sorts

Sexual selection in the genealogical hierarchy provides the following pairings to the other three hierarchies (Figure 5.2b):

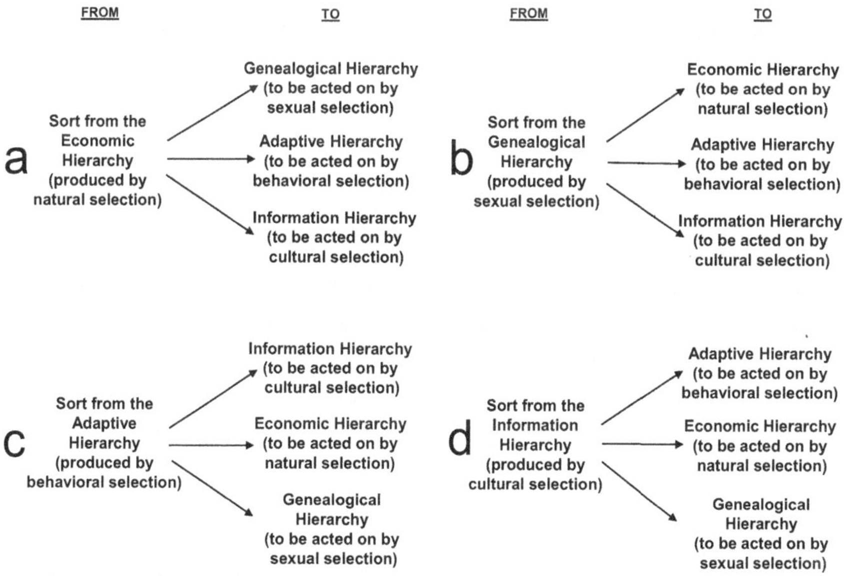

Fig. 5.2 The various sorts arranged by their source

SSs→NS: the *s*exual *s*election *s*ort from the genealogical hierarchy being made available to *n*atural *s*election in the economic hierarchy

SSs→BS: the *s*exual *s*election *s*ort from the genealogical hierarchy being made available to *b*ehavioral *s*election in the adaptive hierarchy

SSs→CS: the *s*exual *s*election *s*ort from the genealogical hierarchy being made available to *c*ultural *s*election in the information hierarchy

As noted by Eldredge (and also Hull, 1980), SSs → NS determines what is available in the way of organic entities and their innate behavioral dispositions for natural selection to act on in the organic economic hierarchy, based on what biologically based attributes sexual selection favors. On the other hand, SSs → BS provides the available range of biological-based physical abilities and behavioral propensities to the adaptive hierarchy, on which behavioral selection can operate by way of decisions concerning choice of physically possible behaviors, biased by any biologically based behavioral predispositions. Lastly, SSs → CS is the differential biological reproductive success of individuals and kin groups in the organic genealogical hierarchy, on which cultural selection acts by way of the differential vertical transmission of propriospects and kin group-associated information systems due to frequency-dependent biases.

Behavioral Selection and its Sorts

Likewise, behavioral selection in the adaptive hierarchy provides the following pairings to the other hierarchies (Figure 5.2c):

BSs→CS: the *b*ehavioral *s*election *s*ort from the adaptive hierarchy being made available to *c*ultural *s*election in the information hierarchy

BSs→NS: the *b*ehavioral *s*election *s*ort from the adaptive hierarchy being made available to *n*atural *s*election in the economic hierarchy

BSs→SS: the *b*ehavioral *s*election *s*ort from the adaptive hierarchy being made available to *s*exual *s*election in the genealogical hierarchy

As noted above, BSs → CS determines what behaviors are available for cultural selection to act on in the information hierarchy—and in the case of novel behaviors, accept as a suitable (or deviant but necessary/acceptable) innovation, or reject outright—based on what thoughtful decision-making (i.e., behavioral selection) makes available for observation and transmission by cultural selection. It is in this process of cultural selection that the various combinations of transmission modes and associated biases explored by Dual Inheritance Theory come into play (e.g., see Boyd & Richerson, 1985; Shennan, 2002). BSs → NS, on the other hand, is the behavioral sort that passes over to the organic economic hierarchy, where it undergoes natural selection on the basis of the various practices' *actual* relative contributions to the fitness of their (organic) practitioners. Simply put, culturally patterned behaviors, whether influenced by behavioral dispositions (the SSs → BS) or not, obviously have consequences for the somatic well-being of the individuals practicing them, and BSs → NS provides the culturally patterned behavioral variability for natural selection to act on. Lastly, BSs → SS is the decision-based sort of culturally patterned reproductive behaviors that are being acted on by sexual selection in the genealogical hierarchy—e.g., success at attracting best/most mates, producing/rearing offspring, etc. (see also Miller, 1999).

Cultural Selection and its Sorts

Finally, cultural selection in the information hierarchy provides the following pairings to the other three hierarchies (Figure 5.2d):

> *CSs→BS*: the *c*ultural *s*election *s*ort from the information hierarchy being made available to *b*ehavioral *s*election in the adaptive hierarchy
>
> *CSs→NS*: the *c*ultural *s*election *s*ort from the information hierarchy being made available to *n*atural *s*election in the economic hierarchy
>
> *CSs→SS*: the *c*ultural *s*election *s*ort from the information hierarchy being made available to *s*exual *s*election in the genealogical hierarchy

CSs → BS determines the range of context-dependent *culturally acceptable* behaviors available for behavioral selection to act on—for people to decide from among to use, or deviate from—in the adaptive hierarchy, based on what cultural selection has made available. CSs → NS, on the other hand, determines the range of contextualized cultural information that flows to the organic economic hierarchy concerning and pertaining to both the social structure/organization of groups in the economic hierarchy and environmental cause-effect relationships relating to somatic well-being. Natural selection then acts on that information in the economic hierarchy without intermediate reference to culturally patterned behavior that may also be based on the same information (e.g., the conceptualized relationship of environmental factors to diseases, as opposed the proper behavioral manner for preventing and treating diseases). Lastly, CSs → SS determines the range of *culturally legitimate* information concerning and pertaining to biological reproduction at the individual and group levels (e.g., conceptions of beauty, understandings about the nature of fertility, the proper social structure and organization of genealogical groups, etc.) available for sexual selection to act on in the organic genealogical hierarchy.

As noted earlier, Rindos (1985: 73) discerned two aspects of these various dynamics in his distinction between the two types of "cultural selection" that he termed "CS1" and "CS2." However, Rindos' CS1 actually alludes to natural selection operating on individuals *with reference* to cultural traits, including artifacts. In other words, it is natural selection operating in the economic hierarchy on both the BSs → NS and CSs → NS sorts. Rindos' CS2, on the other hand, is a reference to what is here being called cultural selection, as it operates in the information hierarchy on the full set of sorts at its disposal (SSs → CS, NSs → CS, BSs → CS).

Smith (2000b: Fig. 2.1) came closer to grasping the full range of systemic dynamics. However, he did not specifically accommodate sexual selection in his model, and more importantly, the dynamic pathways are unidirectional and circular, rather than bidirectional and fully interconnected. As noted above, my own initial attempt to address these dynamics (Rosenberg, 1994) failed by stopping at the replicator-interactor parallels between the organic and metaorganic hierarchies, and even those parallels were not conceptualized correctly.

Reinforcing Sorts, Conflicting Sorts, and Culture Change

The various paired dynamics enumerated above are a description of the individual relationships in which each of the various hierarchies stands to each of the others. In practice, all four modes of selection are operating simultaneously in their respective hierarchies. Thus, while any schematic representation appears static, in operation, their dynamics are anything but static. More importantly, the relative leverage of the various selective forces in matters of culture change is variable and context dependent.

To begin with, when all four are considered in unison, it becomes clear that the adaptive hierarchy is arguably central to their unified dynamics. That is, not only is the adaptive hierarchy the only hierarchy that is both metaorganic and materialistic, making its sort the metaorganic sort most directly visible to the selective forces operating in the organic hierarchies, it is also the most *potentially* flexible of the materialistic hierarchies, being rooted in considered, agential, situationally adaptive decision-making (i.e., human intelligence). Thus, it is the hierarchy most profoundly and immediately affected by the sorts arriving from the other three and the hierarchy that can most variably affect what is potentially available to the other three via its sort.

However, all else being equal, cultural selection, constrained by the Baupläne of the entities in the information hierarchy as well as their existing systemic structure, is by its very nature more conservative than not and tends to act to discourage behavior that departs from existing norms. That is, just as claimed to be the case in biological evolution by the macroevolutionists, the mere existence of a Bauplan tends to promote stasis. Humans are highly social animals and arguably the most group-selected species on the planet. As a feature of the ages-old sorts from the organic hierarchies, we are psychologically programmed to find security in social groups and to want fit into our social groups, most obviously by conforming our behaviors to the degree possible to their behavioral expectations, hence the profound power of peer pressure on us all. Deviation from a given group's norms tends to be penalized, generating very mild social disapproval or ridicule (e.g., teasing) at best and social ostracism or material punishment at worst. Thus, while we may attempt to self-interestedly push at boundaries, without sufficient reason to do otherwise, we tend to avoid outrightly crossing them, and even if we do not fully conform in our beliefs to the information systems carried by our groups, we at least sufficiently conform

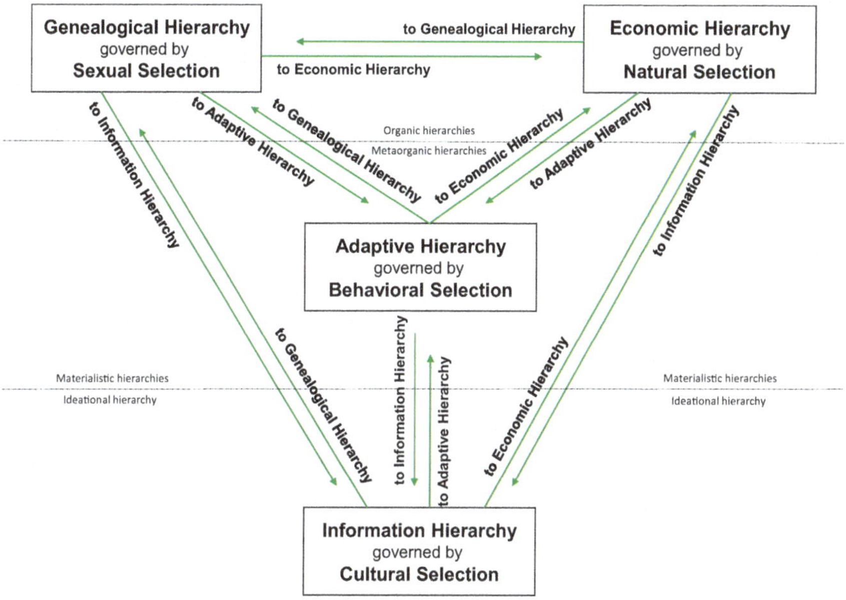

Fig. 5.3 The hierarchical dynamics of microevolution

phenomenologically to the behavioral expectations that stem from those information systems to remain members. How forcefully cultural selection operates to discourage change varies. The more central the beliefs and values being violated by a given deviant behavior are to the structural integrity of the information system, the more serious such penalties will be. Likewise, the more disruptive to the operation of the economic and reproductive systems a given deviant behavior is, the more serious such penalties will be.

In contrast, since they affect existential matters most directly, the sorts from sexual and natural selection tend to foster behavioral change as individuals seek innovative (cooperatively or individually) self-interested ways to improve their somatic well-being and/or reproductive success. Such self-interested behaviors can enhance biological fitness in a culturally approved manner (e.g., see Durham, 1991; Richerson & Boyd, 1999), in which case all the sorts are working toward promoting behavioral ends in conformity with the Bauplan, if not necessarily to the same degree, and there is no significant conflict between the sorts feeding into the adaptive hierarchy and the behavioral selection operating therein (Fig. 5.3). The constraints of rigidity aside, the result is microevolution that proceeds more or less smoothly within the constraints of the Bauplan.

But such a scenario for microevolution presumes that the selective pressures in the three hierarchies producing the sorts feeding into the adaptive hierarchy, and there influencing behavioral selection, are always in balance, do not fluctuate, and promote the same behaviors in the adaptive hierarchy. That is not necessarily the case. Specifically, these other three sorts do not necessarily work toward promoting the same behavioral ends at any given point in time. In other words, they can be working toward promoting conflicting behavioral ends. That is, people will often enough tend to push the ideational boundaries on behavior by behaviorally "gaming" the social, behavioral, and ideational systems to try to gain a relative advantage[2] or perhaps simply feel compelled to engage in unacceptable behaviors for any one of many reasons. When behavioral boundaries are crossed, social sanctions work to discourage further transgressions, which may or may not then persist based on the perceived costs and benefits (correctly calculated or not) of the behavior to the agent.

One can easily envision a situation where natural and/or sexual selection, operating with reference to economic and/or reproductive behaviors and manifest in the NSs → BS and/or SSs → BS sorts flowing into the adaptive hierarchy from the two organic hierarchies, are in opposition to the CSs → BS sort, as perhaps due to a perceived lack of utility in a given context of the culturally approved behaviors for that context. The stronger cultural selection is relative to natural and sexual selection, the more effectively cultural selection constrains behavioral selection's responses to those latter inputs to just what is deemed acceptable within the framework of that Bauplan (Fig. 5.4), maintaining systemic stasis and relatively ineffectual[3] microevolution despite environmental pressures to deviate from approved behaviors in some more efficacious way. This is because the social penalties for unacceptable behaviors are seen as outweighing the value of potential somatic and reproductive benefits to an entity contemplating such unacceptable actions.

The importance of absolute and universal celibacy for the Shakers[4] is a prime example of such a *complete* domination by the sort from cultural selection of the conflicting sort from sexual selection in the genealogical hierarchy. This domination persisted despite its increasingly obvious negative consequences for the organic entities that lived by that information system and thus for the information system itself, which was dying out along with the group that carried it (the group had for all practical purposes died out by the end of the twentieth century, with only three members left alive and the abandonment of all efforts to recruit new members). It should be noted that the Shakers' "demise" was not sudden but gradual, playing out for half a century and leaving ample time for conceptual and/or behavioral changes to be instituted that would have prevented said demise. The fact that none were

[2]The need to promote cooperation and manage competition to prevent it from undermining cooperation is a major concern in all cultural and culturally structured systems. This is because humans are social animals and social animals tend to naturally be socially competitive (e.g., see Boehm, 1999) particularly as a result of sexual selection. That is, higher social standing, by any given yardstick, tends to yield economic and reproductive advantages in attracting most/best mates and producing/rearing offspring. Such competition can easily exist in the context of otherwise cooperative relationships; witness the mid-twentieth-century Hollywood trope of the hero and his closest friend competing for the affections of a woman—the friend usually conveniently and heroically dying at some point to resolve their conflict of interests.

[3]Relative to the forces promoting the deviant behaviors

[4]The United Society of Believers in Christ's Second Appearing, often called the Shaking Quakers or Shakers for the fervent behavior of the participants in their worship services

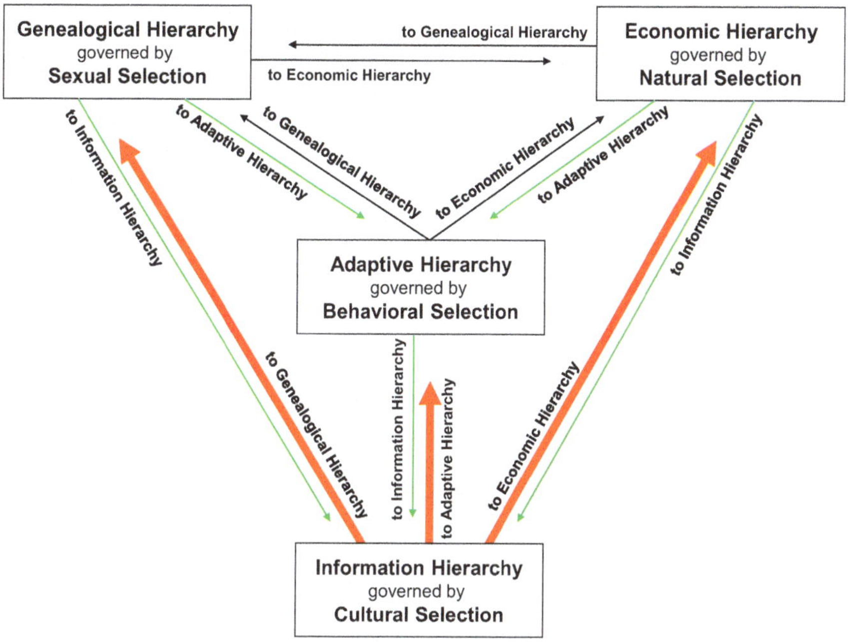

Fig. 5.4 Cultural selection strongly promoting systemic stasis by overpowering environmental forces promoting contracultural change

instituted speaks loudly to the power of the sort produced by cultural selection flowing to the genealogical hierarchy. On the other hand, since natural selection and sexual selection have fundamentally existential effects on the organisms that carry the metaorganic systems, not usually shared by cultural selection, the sorts coming to the adaptive hierarchy from the organic hierarchies certainly have the potential to overpower the sort coming into it from cultural selection (Fig. 5.5). That means that behaviors can, and often enough will, potentially be persisted in despite the imposition of social sanctions, and the more such behaviors are seen as counteracting an existential crisis and the more immediate and/or serious that crisis is, the more likely they are to overpower negative cultural selection in the matter of behavioral selection. To the degree that such persisted-in behaviors violate the Bauplan of the information system, they have the very real potential to cause its rapid collapse. I will return to the specifics of this latter type scenario in the next chapter.

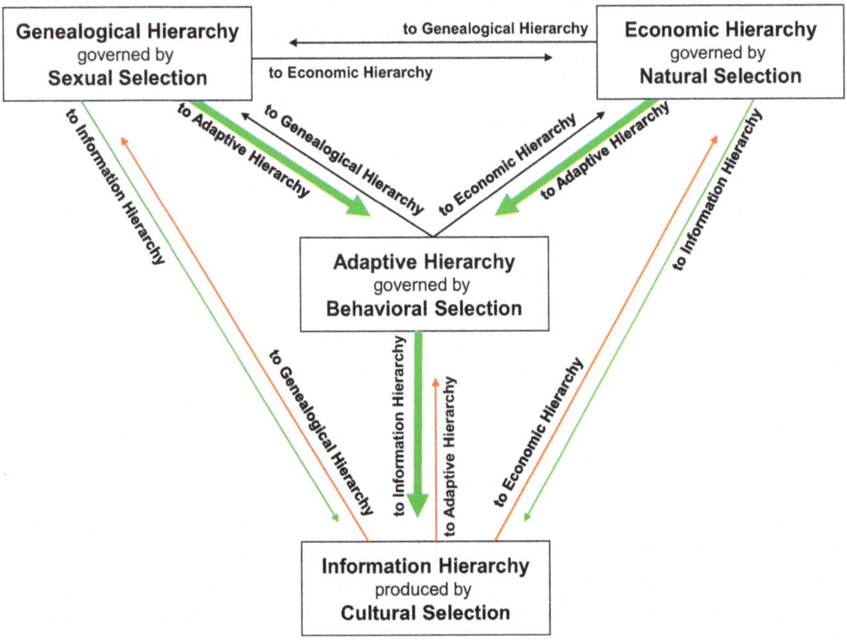

Fig. 5.5 Environmental forces promoting systemic dis-integration

Metaorganic Punctuated Equilibria Require Metaorganic Beginnings and Ends

Evolutionary Culture Theory justifies the limited application of biological evolutionary theory to cultural evolution on three primary counts. The first is that cultures, like species, evolve. The second is that cultures, like species, are related cladogenically. The third is the clear parallels (but not identity) in the informational-phenomenological structure of both organic life and culture. Thus, processes at work in the biological domain, but not logically restricted to it (e.g., selection), can logically be expected to also be at work in the metaorganic domain. Conversely, demonstrable limits to those processes in one domain can also be expected to potentially exist in the other.

For these reasons, it is not surprising that the discussion among biological evolutionists about punctuated equilibria rapidly engendered some limited discussion regarding the applicability of punctuational theory to the evolution of culture (Pro: Diener, 1980; Dunnell, 1980; Gould, 1989a; Eldredge, 1989; Rosenberg, 1994, 2009; Prentiss, 2009; Prentiss & Chatters, 2003. Con: Cachel, 1989). For the moment, Durham's (1990: 196) conclusion—that both "gradual" and punctuated change are possible and that evidence for each seems indicated in the limited available data—still remains the cautious consensus. However, if so, the proper question is not whether both can or do occur. The proper questions are as follows:

Why and when will one or the other occur? What are the processes that produce punctuational change in place of incremental (i.e., microevolutionary) change? What are the respective roles of both types of change in the evolution of cultures and their behavioral and social products?

For biologists, there remains the issue of what exactly are the mechanisms capable of producing the requisite rapid and discontinuous biological reorganizations. However, that question is extraneous to this discussion, which deals with the evolution of culture. This is because diverse aspects of ethnological theory already contain the theoretical basis for a tendency to stasis. They also contain mechanisms that can rapidly produce radically *new* systems independently of selection, and still other mechanisms capable of producing "cascading effects thereafter." In other words, the mechanisms capable of producing both stasis and punctuational episodes of cultural evolution have already been noted by others or are implicit in aspects of culture theory.

Entities come into existence as discreet things, they then exist as discrete things for some length of time, and then at some point in time, they cease to exist; abstractions do not. For the entities in the organic hierarchies, the birth and death of individuals are quite literally just that. For social groups, their "birth" is a social phenomenon that occurs when the first two or more individuals come together as a social group—typically defined for humans as interacting regularly while sharing a consciousness of membership. The "death" of a social group, however, can be either gradual or sudden. The first is when the second to last member, following the departure of others, literally dies or simply drifts away. The second is when, for whatever reason, the social bonds that hold the group together dissolve and the group "dies" in the context of a single event, with the constituent individuals either dying en masse (e.g., the Jonestown mass suicide) or so to speak "agreeing" to dissolve their group (e.g., a corporate bankruptcy or merger). In either case, the group ceases to exist when the last and second to last members part company, as per Eldredge's sloshing bucket, and the group is consequently no more.

Entities in the metaorganic hierarchies, being informational or behavioral features of organic entities, come into being and die with the organic entities that carry them. But they can "die" in yet a third way as well. They can be abandoned by the organic entities that carry them, as in the case of a philosophical conversion, even though those organic entities will then continue to exist and this is arguably the most common way in which they die.

Between Beginnings and Ends there Is Stasis

In between an entity's birth and death, punctuated equilibria involve stasis. Cultural stasis is not the absence of evolutionary culture change. Selection of all types is constant and ongoing; as long as there is variability and competition, evolution is a constant whatever its pace. Behavioral change is by definition deviant with respect to existing expectations. If it is triggered by integrative changes to the information system resulting from previous changes or logical extrapolation from the existing information system, it will always be consistent with the Bauplan of the information system. Deviant behaviors triggered in this manner are at least capable of being accommodated by the existing information system because they do not violate the structural integrity of the Bauplan by virtue of being consistent with it. While not framed specifically in these terms, this idea goes back at least as far as A.R Radcliffe-Brown's formulation of functional (social) coadaptation—that in order to "survive," practices must adapt to other practices (Radcliffe-Brown, 1957). For that reason, resistance to such deviant behaviors is not automatically absolute, and the behavioral changes may eventually produce changes in the relevant information system by sequentially incremental transmissional dynamics—microevolution—as the system attempts to integrate the persisted-in innovative behaviors through cultural selection. In other words, stasis is incremental evolutionary change—microevolution, within the constraints of the Bauplan, which, by virtue of being the structural core of the system, is what does not fundamentally change for the life of the system.

One brief example should suffice to illustrate this capacity for culture to evolve substantially while also maintaining stasis at the level of Bauplan. The overarching American eidos and ethos include a variety of intertwined themes. As ideals, these include limited socioeconomic and sociopolitical egalitarianism, in the form of equal opportunity, equal political say (one person, one vote), equal rights under the law, etc., all intertwined with individualism, the work ethic, etc. (see DuBois, 1955; Hsu, 1961; Slater, 1970). All have been a feature of (the ultimately dominant northern variant of) overarching American culture from the very early colonial period (e.g., see Weber, 1904). In the specific case of limited egalitarianism, the conceptual category specifying who in principle merits inclusion as an "equal" has incrementally evolved from (broadly speaking) all landowning white men (formally established by the founding fathers), to all white men (under Jackson), to all men (beginning under Lincoln), to all men and women (beginning under Wilson).[5] All these changes were latent in the Bauplan, based simply on whether one chooses to understand "men" (in ". . . all men are created equal") as literally males or as humans ("one small step for man"). This is not to imply that these conceptual changes came easily; there were ideological barriers, in the form of since-discredited theories concerning racial and gender differences, which had to be surmounted along the way. Nor is it to imply they have yet been fully surmounted. But behaviors have changed accordingly over time as the sort from the information hierarchy fed into the other hierarchies and the entities in the other hierarchies restructured to reintegrate around the informational/conceptual changes. The sorts from the other hierarchies, in turn, have fed back into the information hierarchy precipitating yet further changes there.

[5] On the other side of that coin, the lingering tensions related to affirmative action (the relative merits of the pro and con positions aside) are, at their most basic level, arguably due to their being seen by their opponents as violating certain socioeconomic aspects of such limited egalitarianism (equal opportunity) and are thus seen as incompatible with elements of the Bauplan and thus "un-American."

The point is that the primacy of ideational or behavioral change in any given specific instance of this process aside, American culture has *changed* significantly in the process; yet these basic egalitarian principles have remained essentially unchanged, as have the other elements of the Bauplan, and in a very real sense so has American culture. But these changes were by incremental cultural dynamics—microevolution—and microevolution does not produce new cultural systems. When the day comes that this particular principle of limited egalitarianism (and the others with which it is intertwined) is no longer a feature of American culture, American culture will no longer exist as the specific entity that it was, regardless of whether what has replaced it is still called "American." What something is called is just a label and has no bearing on what it is; just look at the diverse things all commonly called culture, which, as noted previously, are all quite different.

The point is that the ideational superstructural principles that constitute the cultural Bauplan, being most central to the system, are the most resistant to change. To the degree that they constrain innovative behaviors, they most strongly act to inhibit behavioral change that most endangers the structural integrity of the system as a whole. With respect to hierarchical dynamics, the result is the observable tendency toward what is here called cultural stasis that has already been noted by cultural anthropologists (e.g., Harris, 1979; Barrett, 1991), archaeologists (e.g., Bar-Yosef & Belfer-Cohen, 1992), and scholars in other disciplines (e.g., Veblen, 1915).

The archaeological literature abounds with references to what, in so many words, is here being called stasis. For example, in just my little research corner of the Old World during the Neolithic, Bar-Yosef and Belfer-Cohen (1992: 40) see the Natufian as characterized by what they call "homeostasis" (i.e., stasis), and the Halaf remains sufficiently unchanged over time for Watkins (1987: 223) to also use the term "homeostasis" in describing the degree of cultural stability exhibited by it over the course of its approximately half-millennium long existence. The historical record also abounds with examples. For example, in ancient Egypt, at a minimum, we have several hundred years of what only be considered cultural stasis during each of the Old, Middle, and New Kingdoms despite dynastic changes within each and arguably even stasis lasting all the way from the Early Dynastic through the Ptolemaic periods. The same can arguably be said for Chinese culture from at least Zhou (if not Shang or Xia) times through to the beginning of the twentieth century despite dynastic changes.

Just to repeat, all this is not to say that a cultural system does not change; it is hard to find a modern anthropologist who does not acknowledge that cultures can and do change, though, as also often noted, the capacity to change hinges on the degree to which an innovation can be integrated into the preexisting system (e.g., Sahlins and Service, 1960; see also Durham, 1990, 1991). A culture can evolve gradually by means of one mode of selection or another, but only in directions permitted by cultural selection, with cultural selection itself most strongly a product of the culture's preexisting Bauplan. It can thus gradually become a changed culture by means of microevolutionary processes. It cannot, however (contra Durham, 1991: 461–2), become a new culture. Cultures tend toward stasis, *yet they are by no means static* and, given the Lamarckian elements in cultural evolution, certainly much less so than their biological analogs.

Clearly, natural and sexual selection both play a role in incremental cultural microevolution (as do other processes such as drift). However, these modes of selection can only act on what behaviors already exist to be selected for or against. To the degree that what behaviors exist is determined by what behaviors cultural selection permits and cultural selection is rooted in a superstructural Bauplan, no form of selection is capable of incrementally producing new Baupläne and thereby new cultures. Just as punctuational theory mechanically requires biological speciation to be governed by processes distinct from those that drive microevolution, the evolution of new cultural systems must, of necessity, be governed by processes distinct from those that produce incremental evolutionary culture change.

The Ubiquity of Discontinuities in the Archaeological Record

Biological innovation is always a product of genetic change, a physical process. As such, biological innovation's processes are invariably constrained by its ties to the mechanics of biological reproduction. Cultural innovation, on the other hand, is a product of ideational and behavioral, rather than physical, processes, and its evolution involves some Lamarckian elements. Thus, a radical reorganization in the "key features" of a culture is not only inherently possible; such radical change, by virtue of the Lamarckian elements, can clearly be quite rapid indeed. Most important of all, the archaeological record abounds with cases that strongly suggest that the appearance of new cultural systems is all too often both very rapid and discontinuous because it is replete with discontinuities between named archaeological complexes, such complexes being generally thought of as proxies for some aspect and level of cultural or culturally structured behavioral or social systems.

It should be stressed that discontinuity *does not mean* that phylogenetic relationships between successive such systems are undiscernible. Such relationships are often quite clear and may be quite pronounced, though often enough the apparent points of relationship are so weak that, were it not for the exceedingly strong circumstantial or biological (e.g., Kemp et al., 2017) evidence supporting the proposed relationship, they would be viewed as inconclusive. It *does mean* that the transition from one archaeological complex to its local successor or neighboring peripatric offshoot was both "archaeologically instantaneous" and clearly involved a radical reorganization of the cultural Bauplan, to the point where in many cases there is still substantial debate as to whether the transition was the product of in-migration or a local culture and associated social group's evolution into what followed them. By "archaeologically instantaneous," I mean occurring within less than a century and typically *much* less—witness the historically documented rapid evolution of the Native American plains way of life with the introduction of the horse into the American Great Plains in the first decades of the eighteenth century.

By way of example, just a very small sample of particularly rapid, discontinuous transitions visible in the archaeological record despite often abundant pertinent data[6] includes the archaic to Hohokam/Mogollon/Anasazi transitions in the American southwest; the formative to Olmec transition on the western Gulf of Mexico coast; the Mississippian emergence in the Mississippi and lower Ohio drainages; the Gravettian to Solutrean and Solutrean to Magdalenian transitions in western Europe; the Starčevo/Körös/Criş to LBK transition in Europe; the abrupt beginning and end of both the Varna and Unetice in Europe; the Geometric Kebaran to Natufian and Natufian to PPNA transitions in the Levant; the transition from the PPNB to the Pottery Neolithic in southwestern Asia, the local or peripatric evolution of the central Anatolian plateau's aceramic; and Halaf and Samarra origins in northern Mesopotamia. For some of these (e.g., the Natufian and Anasazi), it has already been claimed that they evolve in the context of punctuational episodes (see Bar-Yosef & Belfer-Cohen, 1992; Berry, 1982). In fact, it is probably not all that excessive to make the claim that at a minimum, rapidity, if not discontinuity, characterizes the emergence of virtually every single *named* complex in the archaeological record. Certainly, something akin to this latter claim has already been made at least with respect to the pattern underlying the emergence of the cultures associated with the sociopolitical systems commonly characterized as state societies (see Spencer, 1987, 1990; Yoffee, 1991).

The point of mentioning these examples is not to argue that in cases of such discontinuities, connections with earlier cultures *do not exist*. Neither is it to argue that transitional sites falling within an archaeological "instant" *cannot* conceivably exist, though by definition such sites must be quite rare. The point is to show that such discontinuities appear to exhibit a consistent pattern of punctuational as opposed to environmentally directed incremental change driven primarily by selective forces. It includes a rapid transition from one cultural system to another; general lack of intermediate forms in individual defining cultural elements undergoing change; and change in sufficient numbers of individual elements as to qualify as a wholesale reorganization (i.e., a new cultural system), not some relatively minor modification of the earlier cultural system.

This pattern, in turn, has three implications for Evolutionary Culture Theory. First, the absence of intermediate forms precludes invoking gradual transformations of customary behaviors (e.g., systems theory, cultural selection) as the mode of change. Second, the rapidity of such transitions arguably precludes the operation of organismic or simple trait selection acting alone to produce the requisite sweeping changes. Even if one invokes the Lamarckian elements that can potentially speed the process up, there is simply not enough time for the feedback that would produce competitive replacement of the requisite number of individual cultural elements (particularly at the superstructural level). Finally, the scope of the documented

[6] Such abundant data would be a potential source of contradictory evidence indicating incremental change and counter the claim that the discontinuities in question are more apparent than real due to the nature of the archaeological record.

changes precludes invoking a sudden transformation in only a few highly visible behaviors to explain the observed changes (e.g., catastrophe theory).

As noted above, there have been occasional acknowledgments that punctuational as well as gradual processes may govern the evolution of culture. As also noted above, there have also been occasional claims that instances of punctuated change are actually evident in the archaeological record. However, in the absence of a clear understanding of punctuational processes and the circumstances under which they reign, archaeologists have generally had only three choices as concerns such apparent discontinuities. The first is to accept such discontinuities as real phenomena but momentarily unexplained as regards the processes that might produce them, a gap this book aims to partially fill. The second is to attribute the discontinuities to new groups arriving on the scene through migration. The problem with this second position generally is that the new culture in question is typically as dramatically different from its (hence often unspecified) spatially removed ancestral culture as it is from its local predecessor and this *spatio*temporal discontinuity invariably remains unexplained. The third is to deny their discontinuous nature and consider such discontinuities to simply be either a product of particularly fast-paced episodes of incremental change; artifacts of the gaps inherent in the (implicitly incremental) archaeological record *that will resolve themselves* if and when more or better data become available (but see the well-documented examples listed above); artifacts of the methodologies employed to study the (implicitly incremental) archaeological record; or a product of diffusion. The fundamental problem with these various denials of discontinuity is that all these views have in common a predication on the premise that a cultural system is essentially unbounded, giving it an undifferentiated potential for change in any direction. That view of cultural systems is false because it does not accord any role whatsoever to the constraints imposed by any given system's existing structure, not to mention the constraints imposed by its Bauplan.

Chapter 6
The Deaths and Births of Cultural Systems

> *"When the gap between ideal and real becomes too wide, the system breaks down."—Barbara W. Tuchman*

An Information System's Capacity for Change Is Not Open-Ended

Within Anthropology and Archaeology, the vast majority of evolutionary, ecological, and materialistic perspectives of all kinds implicitly treat cultures as completely open-ended systems. By that I mean that whether any given explanation invokes selection or transformation, nothing is ever seen to stand in the way of selection for or transformation to the cultural state for which an explanation is being proposed. But, as was noted by Diener (1980), culture arguably has considerably less potential for incremental adaptive change than commonly thought. The erroneous view that it does is an outgrowth of the first's foundation in biological microevolutionary theory and its incrementalism, and the latter two's foundation in cultural and ecological systems theory. As such, these perspectives implicitly view cultures as having a virtually unlimited potential for incremental adjustment to changing external and internal variables (e.g., see Kirch, 1984; Bettinger, 1991). Some capacity for incremental coherent internal modification in response to external factors certainly exists in the form of microevolution. However, the presumed scope of this capacity to produce entirely new cultural systems is highly questionable. That is, such an unlimited capacity for gradual change requires the nonexistence of such a thing as negative cultural selection feeding into the adaptive hierarchy and a view of cultural malleability so caricatured as to be scarcely believable when compared to observable reality. Alternately, it requires the ruthlessly efficient operation of natural and sexual selection on the organic entities in question with individual reference to any and all individual cultural traits. As suggested by Gould, faith in such "vulgar Darwinism" (1983: 369) is unwarranted, even in biological evolution. In other words, culture may be more of a closed system than how it is usually treated.

Aside from the constraints imposed by its Bauplan on an information system's capacity for change, there is also the matter of increasing systemic complexity in the

M. Rosenberg, *The Dynamics of Cultural Evolution*, Studies in Human Ecology and Adaptation 12, https://doi.org/10.1007/978-3-031-04863-0_6

two social hierarchies and the informational one constraining further change into only compatible channels, which become increasingly narrow as complexity grows. Systemic complexity in these can arise by various means, not the least of which may be biological selection (on organic groups and thus the SSs → CS and NSs → CS sorts from the organic hierarchies going to the information hierarchy) for the relative operational refinement of complex social systems over simpler ones, and also their ability to hold larger groups together on a permanent basis, there often being strength in numbers in the context of intergroup competition. For both social and informational entities, another obvious source of complexity is the assembly of cultural and culturally structured social "contraptions"—the innovative, though not necessarily efficient, combination of existing elements, which by virtue of their preexistence are congruous with the existing Bauplan and thus available to meet perceived needs (e.g., see Renfrew, 1978). As the systemic complexity of the system thus grows, the greater grow the number of collateral elements and more importantly the greater grows their interconnectedness, producing ever more structural rigidity. Add to that structural rigidity the operation of human agency, in that individuals and groups will tend to act to protect their interests as they exist in the context of the then-current state of the system. The greater the system's complexity, the more numerous such interests purposefully being furthered by thwarting any changes to the system that would harm them.

If the value of novelties is culturally defined at least in part in terms of their congruity with these collateral elements and interests, then it follows that an innovation's chance of acceptance hinges at least in part on the likelihood of it achieving such congruity. If a cultural system gravitates toward internal integration, aside from the constraints imposed by the Bauplan, it logically follows that innovation is also highly constrained by any at-that-moment preexisting interconnected collateral elements and interests. It further follows that the greater the complexity of a system and its interconnectedness, the stronger these structural constraints on further change to that system outside of acceptable channels. Thus, the constraints of Bauplan aside, a system's susceptibility to internally generated incremental change tends to vary inversely with its complexity.

The Death of Cultural Systems

New cultural systems are born in one of two contexts, but both involve essentially the same process, one that is both very rapid and discontinuous. The first context is one wherein the emergent system is born as the product of peripatric fissioning from an existing system, as would occur for example in the context of out-migration and colonization. The second context is the phoenix-like emergence of a new system from the rubble of a cultural system or subsystem that has collapsed and been abandoned by the organic entity that carried it. Since the death of a cultural system in this second context is integral to the birth of what potentially replaces it, it seems appropriate to reverse the intuitively logical order of addressing births before deaths, and start with deaths.

Selective forces can only operate on what exists and do operate on what does exist, which in anything but the very simplest systems requires them to operate on the phenomenological expressions of combinations of (interconnected) elements, be they genetic or cultural traits. The greater the number of collateral elements, the greater grows the potential for suboptimal collateral elements (as long as they are not directly lethal to the system) to hamper the system's functional efficacy while being directly hidden from selective forces due to their increasingly indirect functional roles and/or their obscurity due to their embeddedness as part of a system. Then, there is also the matter of self-interested behaviors on the part of individuals and groups. Once a system exists, it will be defended by individuals and groups who benefit from its existence and thus constitute vested interests, even as the system's functional efficacy is increasingly weighed down by the addition of suboptimal collateral elements. On balance, such vested interests resist changes that jeopardize those interests and constrain further change to only those directions that minimize the jeopardy to those interests.[1] Thus, as complexity grows, selective forces stagnate and change grows to be increasingly dominated by nonselective drift, an aspect of microevolution. In other words, over time systems become increasingly rigid and increasingly incapable of radical change, even within the constraints of the Bauplan, thus reinforcing stasis.

While deviant behaviors consistent with the Bauplan of the information system, and thus potentially capable of being accommodated by it, will (if adopted by others) produce incremental microevolutionary change, deviant behaviors are not of necessity always congruent with the Bauplan and in such cases are incapable of being accommodated by it. As noted previously, not all sorts are equally powerful at any given point in time and not all sorts are necessarily working toward promoting the same behavioral ends. As also noted earlier, the more central given beliefs and values are to the structural integrity of an information system, the more serious will be the social penalties for engaging in a given deviant behavior that violates them and the more serious the degree to which such behavior does so, the more serious the penalties. To that, we can add that the more disruptive to the operation of the economic and reproductive systems in particular a given deviant behavior is, the more serious such penalties will be (Fig. 5.3). Thus, under normal circumstances, in cases of these sorts being in opposition, the CSs → BS sort will tend to dominate the SSs → BS and NSs → BS sorts in matters of behavioral selection, thereby also promoting stasis.

[1] As an example from recent history, one has only to look at the history of largely unsuccessful attempts to restructure, or even just to reform the American public school system over the past 60 years to clearly see this inertia in operation. As an example from Medieval history, one has only to look at the succession of fourteenth-century popes who tried *without success* to reform the church's financial structure, set in place with the move of the papacy to Avignon in 1309, in an effort to correct the abuses that would ultimately precipitate the full-blown crisis of The Reformation in the early sixteenth century (e.g., see Tuchman, 1978).

However, as also noted earlier, natural selection and sexual selection have existential effects on the organic entities that carry the metaorganic systems, only indirectly shared by cultural selection. Moreover, they are often more directly, even if not always more clearly perceptible, even when perceived through a cultural filter. For example, death is experienced directly whether ascribed to malnutrition, violence, magic, or the actions of supernatural powers. Thus, given a situation wherein the behavioral path to existential well-being will lead to social disapproval and the behavioral path to social approval to existential harm, the sorts coming into the adaptive hierarchy from the organic hierarchies do have the potential to overpower the sort coming from cultural selection as regards behavioral selection in the adaptive hierarchy and produce innovative *contra*-cultural behaviors that promote existential well-being at the expense of the social acceptance that would potentially come from more culturally appropriate but less efficacious behaviors.

Phrased in more formal terms, in the adaptive hierarchy, the NSs → BS and SSs → BS sorts have the potential to trigger calculated, innovative behavioral change opposed by the information system's Bauplan and its expression in the CSs → BS sort (Fig. 5.5). Any such innovative agential behaviors generated by the relatively stronger NSs → BS and SSs → BS sorts will naturally draw social sanctions aimed at discouraging them. But, the more such potential contra-cultural behaviors are seen as efficacious and the more they are seen as counteracting an ongoing or imminent existential harm, the more any such contra-cultural behaviors will potentially not just be persisted in, but spread despite the social sanctions they generate. Moreover, the more critical that potential harm, existential or otherwise, the more likely the sorts from any or both the organic hierarchies to the adaptive hierarchy are to overpower the sort provided by cultural selection to the adaptive hierarchy, the more likely such contra-cultural behaviors are to be behaviorally selected for, and the more likely such behaviors will be persisted in even in the face of severe social sanctions.

In plain English, if the need is seen as great enough, individuals and groups will *tend* to engage in behaviors that promote short term (i.e., somatic) and long-term (reproductive) survival at the expense of "behaving in accordance with cultural expectations," and the greater is seen the threat to survival, the more likely they are to be so engaged in. All this is not to say that the CSs → BS sort will always in such circumstances be overridden by the NSs → BS and SSs → BS sorts, in which cases natural and sexual selection will act to eliminate the information system by operating directly on the organic carriers of that information system. The Shaker example noted previously is a prime example of this. As evident in their formal name, they believed that the second coming of Christ was imminent and that, in that context, their spiritual "survival" simply took precedence over the long-term earthly survival of the group, which for all practical purposes[2] did not in fact survive, conjointly with the information system that the group carried.

[2]The group's "death" as a social group and that of the group's information system is imminent, with only two members aged 60 and 78 left alive as of January 2017.

However, in cases where the sorts from the organic hierarchies do override the CSs → BS sort, persistence in such contra-cultural behaviors will cause the information system to collapse because its rigidity—a product of Bauplan plus any evolved complexity and associated interest groups' resistance to change—makes it incapable of accommodating the innovative behaviors. When enough individuals cross some behavioral threshold, policing the violated old norm(s) becomes impossible and the conceptual "girder" underlying the norm(s) buckles, setting off a chain reaction of cascading effects in which any other interconnected "girders" then also buckle and the entire system or subsystem ultimately *dis*integrates. It is then rapidly replaced by an emergent new system that crystallizes around accommodating the innovations that precipitated the collapse of the old system, which were behavioral responses (though not necessarily optimal or even necessarily efficacious ones) to selective pressures generated by competition in one or both of the organic hierarchies.

To phrase all this in still another way, this time in the terminology of cultural materialism as opposed to that of evolutionary culture theory: stress which is experienced at the infrastructural level has the *greatest* potential to generate deviant behavioral innovations that people will persist in *despite* any threat of negative sanctions. In such instances, potential negative cultural selection is countered by perceived self-interest in the most critical of all matters—survival. Stress can also generate innovations at the level of structure and superstructure. However, to the degree that stress is perceived in other than infrastructural terms, self-interest is more likely to revolve around socially or culturally defined goals rather than simple survival. Consequently, perceived benefits are less likely to outweigh negative sanctions. Thus, the more radically system-threatening a contra-cultural structural or superstructural level deviant behavior is, the less likely it is to become established in the face of cultural selection. The higher in the materialist hierarchy of norms a deviant behavior is contemplated, the less powerful is the incentive to radically innovate in a disapproved manner by anyone who does not have a degree of immunity to the potential sanctions.[3]

To the degree that any infrastructural level deviant behaviors actually have selective value and contribute to the fitness of those practicing them, they will be favored by organic modes of selection. More immediately important, to the degree that such innovative behaviors yield what are seen as perceptible net benefits, they will be emulated by others and thus spread by biased transmission (see Boyd & Richerson, 1985). Conventional views of cultural materialistic relationships hold

[3]For example, it took a Roman Emperor (Constantine) to establish Christianity as the culture's dominant religious ideology approximately 300 years after its emergence as a significant movement within the Roman Empire and during which time its adherents were often enough persecuted. The same leadership-directed change was commonly at the center of the conversion to Christianity of Europe's other pagan groups. In contrast, the Cathars and Protestant Reformation were never legitimized by the highest levels of the Church's leadership, resulting in the Albigensian Crusade in the case of the former and the violence (e.g., St. Bartholomew's Day Massacre) culminating in the 30 Years War in the case of the latter.

that at some critical threshold this eventually produces a sequential reorganization of first structure and then later superstructure. This occurs as the cultural system gravitates toward reintegrating around these necessity-based innovative infrastructural behaviors. To the degree that an innovation is in conformity with key superstructural elements of the Bauplan, or can be rationalized by extension of key superstructural principles (e.g., all "men" are created equal), reintegration is possible—the system is potentially open and this process of reintegration and incorporation (in conjunction with selection) can occur. The Industrial Revolution in northern Europe and North America is a historic example of such an accommodatable infrastructural innovation and the invention/introduction of the plow at the end of the Chalcolithic in southwestern Asian food-producing systems may very well be a prehistoric example of one. Competition-induced stresses are not a prerequisite for the inception or establishment of such innovations (though they certainly can be) and in such cases the result will be incremental microevolution. It should be emphasized again that microevolutionary change does not necessarily equate with minor change, nor is it necessarily limited to just one reintegrative cycle. Logically, any superstructural changes resulting from lower level changes will likely alter at least some aspects of cultural selection. This may then make *possible* the acceptance of a whole new range of infrastructural, structural, and superstructural innovations, with this incremental process continuing until the potential for change of a given Bauplan theoretically becomes exhausted.

However, to the degree that a possible deviant behavior is contra-cultural, the cultural (i.e., informational) system is incapable of accommodating it and microevolution in that direction is foreclosed. Moreover, to the degree that such innovative contra-cultural behaviors are nevertheless engaged in by an ever-larger number of individuals, the greater are the contradiction-induced internal stresses that build within the system. In the case of contra-cultural behaviors at the level of infrastructure, this does not generate a relatively smooth *re*integration of socioeconomic and political structures around the new expedience-based infrastructural behavioral norms. Instead, I suggest that the resulting stresses will produce the *dis*integration of these structures followed hard by key elements of superstructure as well. To the degree that contra-cultural behavior is not just different, but conflicts with both the ideological underpinnings of the larger informational system and/or the operation of the behavioral system, infrastructural behavioral innovation has the potential to quite literally destroy the entire system as a system. In so doing, it will initiate a punctuational episode. Contra-cultural innovations at the level of structure and superstructure have respectively more limited capacities to precipitate punctuational episodes as all-encompassing as ones precipitated at the infrastructural level.

It should be noted that while it is the information system that collapses as a result of contra-cultural behaviors, the punctuational episode does not just result in a new information system having a different "organic design" than the old one, as focused on thus far; that is just what is going on at the heart of the process. It will also be expressed as a new "organic" design of the behavioral system (e.g., see Prentiss & Chatters, 2003; Chatters & Prentiss, 2005; Zeder, 2009), as well as by a new "organic design" of the social systems (e.g., Spencer, 1997) structured by both the new information system and new behavioral system, for without corresponding new behaviors the new information or social systems would not replicate,. This is because evolutionary culture change must be expressed as all of these, whether we are talking about macro- or microevolution, or even drift.

It should also be noted that such an essentially materialistic view of punctuational change is not incompatible with Wallace's (1972) essentially ideational and punctuational concept of "paradigmatic processes" modeled after Kuhn's (1962) paradigmatic revolutions. Wallace, following Kuhn, focused on the collapse (i.e., disintegration) of superstructural (i.e., belief) systems due to a loss of faith in their validity, resulting from their perceived incompatibility with new realities. All paradigmatic revolutions are by definition conceptual. *That does not, however, confine the source of all such occurrences to just superstructure.* At the superstructural level, a paradigmatic revolution results from loss of faith in the conceptual underpinnings of the system (e.g., the belief system). At the structural level, it results from loss of faith in the efficacious workings of the system (e.g., the social, political, or economic systems), and at the infrastructural level it results from loss of faith in the productive or reproductive capacity of the system (e.g., the subsistence and/or reproductive techniques and technologies).

In the case of superstructural revolutions, it results in the immediate substitution of one conceptual system for another (e.g., Christianity for Paganism in the Roman Empire). However, a loss of faith in the workings or productive capacity of the system is quite possible in the face of continued faith for a time in the ideological rationale for the system. Witness the lingering faith in altruistic communist principles by many Russians, as expressed in nostalgia, even after their faith in the workings of the soviet system was long abandoned by most. Structural and infrastructural revolutions are not produced through superstructurally rationalized innovation, but through innovative behaviors that are incompatible with still held beliefs. This perforce *leads* to the ultimate collapse of the old beliefs and only later to the establishment of a new superstructural system. Even though it is the new "faith" (i.e., paradigm) that ultimately defines the Bauplan of the new information system (i.e., culture), paradigmatic revolutions result *from a loss of "faith,"* not from the *establishment* of a new "faith." Establishment follows loss, for if the old faith had been maintained there would be no basis for establishing a new one.

Two examples should suffice to illustrate this process, one historical and the other ethnographic. The historic example is the collapse of the Soviet Union as a societal level sociopolitical/socioeconomic (i.e., structural) system in the face of socioeconomic reforms that were triggered by the existing system's inability to compete economically, and therefore militarily. These structural reforms, labeled *perestroika*, were incapable of being accommodated by the indiscriminately altruistic superstructural eidos (e.g., from each according to their ability, to each according to their need) and ethos (e.g., self-interested economic behavior is anathema) of the existing Marxist-Leninist communist system and the result was that both the structural and superstructural levels of that system rapidly collapsed to be replaced over roughly the next two decades first by a transient oligarchic kleptocracy and ultimately by the nationalistic (as opposed to internationalist) ex-KGB thugocracy that emerged from the rubble of the old system (see Belton, 2020 for a detailed account) and which accommodated self-interested economic behavior on the part of both individuals and groups.

The second, ethnographic example was provided by Yellen (1985: 47–48), who, in passing, briefly recounted the consequences stemming from a San man named Rakudu's choice of contra-cultural economic actions. It is an interesting case in point that at once illustrates the power of the CSs → BS sort, while simultaneously suggesting precisely the proposed outcome of a persistent contra-cultural infrastructural innovation. Rakudu and his family, for unspecified reasons, had decided to produce rather than forage for at least some of their food—specifically by maintaining goats. The resulting problems arose from the fact that food production requires some level of resource ownership to assure both a return on invested labor and the preservation of capital (in this case, breeding stock) essential to a delayed return system. This, however, is antithetical to a code of conduct typical of immediate return systems that stresses generalized reciprocity (e.g., see Bird-David, 1992; Boehm, 1999), particularly in the form of unrestricted sharing of meat. Thus, Rakudu's behavior led to the widespread view that he, and by extension his sons also, were stingy and to the application of subtle social sanctions.

To begin with, Rakudu's case illustrates an important point, to the effect that systemic disintegration can occur to any order system from the individual's propriospect on up, as well as at any materialistic level. More importantly for present purposes, the latent hostility generated by Rakudu's actions (see also Yellen, 1990), and most clearly embodied in the difficulty his sons had in both getting and subsequently keeping wives, forced him to abandon his innovative subsistence behaviors. Moreover, there is ample reason to think that such social dynamics are not unique to just this case (see Chase, 1989). The point is, suppose that subsistence-related stresses of one sort or another had fostered the persistence in such behaviors by an ever-larger number of hunter-gatherer individuals (e.g., see Rosenberg, 1990, 1998) who viewed the source of those economic stresses as a more immediate concern than the social sanctions levied against them as a result of their actions. The persistent latent, if not overt hostility arising from the resulting clash between emergent necessity-based subsistence-oriented behavioral norms (having selective value in the economic hierarchy) and the Bauplan of the preexisting system would necessarily have come to permeate the social domain. This latent hostility, I suggest, would surely have caused the *rapid disintegration* of the full range of socioeconomic and political structures integral to a mobile foraging, immediate return lifeway, the eventual abandonment of the beliefs and values that originally motivated the behaviors characteristic of the previous lifeway, and their replacement by a new system capable of accommodating the new infrastructural behaviors and their social consequences.

For example, given the localized variability in resource availability that characterizes any given landscape, it is reasonable to assume that any stresses experienced by social entities in the economic hierarchy would be experienced, at least initially, by some localized communities and not by others. Thus, if one or more local communities of mobile foragers experiencing such stress opted to go a variant of

Rakudu's innovative route and produce food, one way to deal with the subsequent social fallout from other groups not innovating for whatever reason would be by also then instituting local-community-level endogamy, as at Hayonim (see Smith, 1973), a Natufian site in the Levant, where "in-breeding" is evident. Nor is Hayonim an isolated case. Similar community-level endogamy has also been noted at the slightly later site of Basta, an aceramic Neolithic community also in the Levant, where it cannot be explained on the basis of either environmental or demographic factors and "appears to have been a deliberate choice" (see Alt et al., 2013; see also Bodet, 2019 for sites in central Anatolia).

The Birth of Cultural Systems

New cultural systems can come into being in either of two basic contexts. The first is the phoenix-like emergence of a new system from the rubble of an old one that has disintegrated, as detailed above; the second is the peripatric out-migration of a breakaway group. While all beginnings short of the big bang are inherently "fuzzy," in both contexts the birth of a new cultural system is rapid. As noted previously, discontinuities between named archaeological entities abound in the archaeological record, with enough such discontinuities not attributable to gaps in the archaeological record (by virtue of both the entities temporally bracketing the gap being well documented) that there is good reason to conclude that such discontinuities are much more generally not the product of gaps in the archaeological record than the specific cases enumerated. And, if the evolution of the American Great Plains historic way of life is any indication, the evolution of new cultural and culturally structured behavioral systems can occur within the approximate span of a generation or two.

Given the disintegration of a cultural system or the out-migration of a breakaway group, what are the forces that shape the new emergent cultural and behavioral systems? Cultural materialism and evolutionary culture theory both accord that role largely to the incremental operation of selective or historical forces of one kind or another. However, both punctuational theory and a number of ethnographically based studies (see Barrett, 1991) suggest that stochastic processes play a prominent, if not the dominant role in the *shaping* of new cultural systems (see also Boyd & Richerson, 1992: 187ff.). This implies that adaptive *maximization* is not a consistent feature of the emergent norms that coalesce to form the new systems. It also implies that what develops is not so much *determined* by the new behavioral norms, as it is at once *permitted* and constrained by them. Thus, in the terminology of cultural materialism, the dynamics of punctuational evolutionary culture change should properly be called infrastructural *opportunism*, not infrastructural determinism.

Elements of Foster's (1960: 227ff.) concept of *cultural crystallization* and its secondary implications dovetail quite nicely with such a view of punctuational change and infrastructural primacy. Foster saw the primary determinant of which Spanish cultural elements were and were not established in post-conquest Latin

America as little more than the *order* of their introduction. He suggested that the Spanish conquest was followed by a relatively short interval during which the institutions of emergent Latin American cultures first developed. Southern Spanish elements introduced during this short "fluid period" were likely to be incorporated on an as-needed, ad hoc basis into the emergent system. Northern Spanish elements introduced for the most part only later, after the new system had already "crystallized," were less likely to take hold proportional to the degree that their functional equivalents were already integral parts of the new system.

Barrett (1991) takes the implications of Foster's case one step further. He suggests that cultural systems in general will tend to what is here being called stasis following a relatively short formative period characterized by the opportunistic incorporation of essential elements, some traditional and others innovative. The specifics of a cultural system are not so much determined by "the greater suitability of one set of customs" over another, as "simply a matter of which practices [come] into use before the system [hardens] into a [functioning] mold" (1991: 118). Thereafter, what Boyd and Richerson (1985) refer to as frequency-dependent processes would insure their continued practice. Anyone who doubts the validity of that generalization has only to look at the QWERTY keyboard layout (see Gould, 1987), the triumph of VHS over Beta-format VCR machines (see Arthur, 1990), and the myriad clearly suboptimal embedded norms in American, or for that matter any other culture.

I do not necessarily see such a process of rapid crystallization as necessarily always producing quite as rigid a product as perhaps does Foster, or even Barrett, though it potentially could. However, those elements of their models that propose rapid cultural coalescence, likely suboptimal functionality of emergent institutions, and the constraints of precedent, are consistent with readily observable reality and thus apparently correct, so far as they go. These have profound ramifications for any evolutionary view that considers organic modes of selection to be the primary force initially shaping the emergence of both cultures and culturally structured behavioral and social systems.

The short formative period during which the Bauplan of the emergent system crystallizes around either the contra-cultural behaviors that precipitated the collapse of the old system or the necessities of the migrated-to environment will, to a significant degree, be characterized by the necessity to provide for all other (than any already met by contra-cultural innovations in cases of collapse) basic needs in the absence of a full panoply of established understandings for how to properly do so. Because all true evolutionary processes are opportunistic, it follows that the emergent system will incorporate both need-based innovative behaviors as well as all *potentially* salvageable (i.e., that do not conflict with emergent necessity-based norms) behavioral and organizational elements of the earlier system(s), the latter for no other reason than their familiarity.[4] In the case of the example mentioned earlier of what emerged from the collapse of the soviet system, the intelligence service, not

[4]See Renfrew (1978) for a more detailed discussion.

having had a direct organizational role in the running of the old failed system and filling a continuing need, was carried over in modified form and, more importantly, remained sufficiently organized to ultimately consolidate autocratic power, though other well-organized economically peripheral carried-over groups, such as the military, theoretically at least could have instead had they been inclined to and/or been more nimble.

This propensity to (all else being equal) carry over nonconflicting elements applies also to what Durham (1991:368ff.) calls "opposed" (i.e., maladaptive) elements, including those "imposed" (1991: 198) on the majority of the group by individuals or groups with the power to do so.[5] Thus, it is theoretically not even necessary for traits to have any net selective value at all for the majority, who are/remain in essence thus parasitized, in order for those traits to become part of an emergent system. All that is necessary is that they not *conflict* conceptually with the key infrastructural and directly associated higher level elements of the emergent (dominant) organic and metaorganic systems.

Moreover, Foster's data (and the evidence from the archaeological record) strongly suggest that the proposed time frame is too short for the effective incremental operation of organic modes of selection in the initial formation of emergent systems. Consequently, barring precognition, the key determinant of incorporation is certainly not long-term adaptive/selective value. Rather, beyond apparent proximate utility to those with the largest role in shaping the emergent system, the key determinant is familiarity and perceived compatibility with those key utilitarian elements of the emergent system, particularly those that precipitated the collapse of the old system, if such a collapse was what precipitated the emergence of the new system.

Gladwin's (1957) study of the historically derived difference between Cheyenne and Comanche courtship practices illustrates this carryover of analogous norms into emergent Plains cultures. Oliver's (1962: 58) observation regarding the historical origin of the differences in sociopolitical organization among the Plains tribes does the same. This is not to say that biological and cultural selection may not later refine the system within the constraints imposed by the Bauplan, but the Bauplan and hence the directions left open for selection are determined by what develops during the process of crystallization. This is also not to say that selection does not subsequently operate on the system as a whole in competition with other systems (see below). It is simply to say that neither selection nor intent to optimize some objectively correct vision of long-term adaptiveness are likely to be significant factors in shaping the system that forms during the crystallization phase and thus in shaping the system's Bauplan. In other words, while the new Bauplan and the system of which it is the core is very much precipitated by the operation of natural

[5]For example, in the case of colonial Latin America, one set of groups was sufficiently centralized and militarily powerful to forcibly "impose" innovations, including carried-over class-based "impositions" on the indigenous populace that constituted the majority of the descendent societies and the cultures they carried (see Wolf, 1982:131ff.).

and/or sexual selection in the organic hierarchies (i.e., SSs → BS and NSs → BS) and not drift, the Bauplan's formation is not actively shaped by those biological forces (i.e., SSs → CS and NSs → CS). Those latter selective forces only come into significant play again once the emergent system has crystallized and it is then that the sorts from the organic hierarchies return to playing crucial roles in the competition between systems at all levels of all the hierarchies.

In cases where new cultural systems are born in the context of migration/colonization by breakaway groups, it should also be borne in mind that such groups are likely relatively small as compared to their parental populations and not at all likely to be fully representative carriers of the entire parental information system. Thus, in such cases, what initially crystallizes as the emergent *new* information system and its Bauplan is significantly shaped by the cultural equivalent of founder's effect and other forms of drift. This often allows for a period of rapid experimentation with adaptive behavioral responses—though again not necessarily optimal ones—to economic and reproductive exigencies characteristic of the breakaway group's new environment that may have been precluded by the old Bauplan prior to the split. This is because the new emergent information system is still fragmentary, in a state of flux, and thus relatively flexible due to a lack of consensus as to what exactly "the rules of the new game" are.

As a result, the new emergent information system carried by breakaway groups has the potential to be more dramatically different than the system it is derived from than does a system that is born out of the collapse of its genealogical antecedent. This is because in the birth of new cultural systems resulting from the collapse of an older system the bulk of the earlier Bauplan is conceptually still theoretically available for carrying over. While those elements of the Bauplan that were opposed to the innovative behaviors that precipitated the collapse of the older system will naturally not be carried over, any other elements that are not opposed to the new Bauplan and that can be integrated into the emergent system stand a good chance of being salvaged from the old system, again whether they are suboptimal or not. In contrast, it is inherent in the small numbers underlying founder's effect that an emergent breakaway system, whether it be an organic or metaorganic entity, will very likely be quantitatively unrepresentative of the parental entity, with the result that there will be fewer elements of the parental system available to potentially carry over into the new emergent system.

Finally, while all new cultural systems are proposed to appear in the context of punctuational episodes, the degree of differentiation between ancestral and descendent cultures is proposed to also be determined by whether punctuated change was precipitated at the infrastructural, structural, or superstructural level. Contra-cultural infrastructural innovation will necessitate (but not dictate the form of) innovative supportive subsystems at all higher levels. Higher level contra-cultural innovation (including sociopolitical revolutions) will likewise precipitate supportive innovations at that and any immediately higher, though not directly at lower, levels. Thus, such innovations will affect fewer subsystems and therefore be less sweeping in their systemic consequences on any higher order entities within their respective organic or metaorganic hierarchy. Moreover, the lower in the materialist hierarchy of norms the precipitating contra-cultural innovation, the more likely it will be that fewer components of the parental system would have perceived functional worth relative to the new norms. Thus, the lower the level of the precipitating contra-cultural innovations, the more different the emergent system would likely be from the system it replaces. For example, the difference between a hunting-gathering culture (e.g., Kebaran) and its food-producing (e.g., Natufian) successor should have been quantitatively and qualitatively greater[6] than that between an early/quasi proto-authoritarian culture (e.g., Halaf) and its nominally egalitarian predecessor (e.g., Hassuna) had that change itself not been precipitated by a previous change at a lower level. The latter difference, in turn, should theoretically have been greater than that between two cultural systems divided by a superstructural level punctuational episode (e.g., Pagan vs. Christian Rome, Coptic vs. Islamic Egypt).

Cultural and Culturally Structured Systems and Group Selection

Cultural systems, be they those carried by individuals or groups of any level, crystallize rapidly in a manner that to a significant degree involves stochastic and selective processes operating in the metaorganic hierarchies rather than selective processes operating in the organic hierarchies. However, once the emergent system has substantially crystallized, selective processes in the organic hierarchies reassert themselves—potentially with a vengeance, with their sorts now feeding existentially relevant input to the selective processes in the metaorganic hierarchies; and, those selective forces operate at all levels of all the hierarchies. Those selective forces are generated by reproductive and economic interactions in the organic hierarchies between humans and the nonhuman environment and by interactions, including competitive interactions, between human entities. Such interhuman competitive interactions can be between individuals and between groups of cooperating individuals, as well as groups of cooperating groups, all of whom can potentially also be either cooperators or competitors in other specific contexts.[7] Moreover, selection can operate simultaneously on multiple levels and not always to the same ends. Those interaction-driven selective forces in the organic hierarchies, as noted above, have the potential to generate sorts that overwhelm cultural selection, precipitating the death of the cultural system that regulated the interactions. At any level above the individual, what we are dealing with here is group selection.

As noted previously (see sect. "Macroevolution and Group Selection"), while within Biology the possibility of group selection, as originally envisioned by Darwin

[6]"Greater" here refers to both the "vertical" range of affected levels within the hierarchy of norms and the "horizontal" range of affected norms within each hierarchical level.

[7]E.g., peasant groups versus nobles within a feudally organized kingdom and the peasants along with the nobility of that kingdom versus the inhabitants of another kingdom.

(1871) in "for-the-good-of-the-group" terms had been pronounced dead at the hands of genetic reductionists by the 1970s, an alternative "top-down" view of group selection was gaining credence by the 1990s. That is, it was demonstrated that the declining intragroup relative frequency of individuals with a given good-for-the-group characteristic (e.g., altruism) produced by organismic selection can be overcome by the relative *inter*group success of demes having such members. Moreover, it was proposed that selection can without question operate to favor exclusively group-level traits in any of the hierarchies. This would include traits that fall under the heading of emergent fitness-enhancing traits, such as a *group's* system's ability to more readily propagate itself than can the others it is in competition with and ones that fall under the heading of emergent traits, such as forms of social structure and/or organization, that enhance the system's success at benefiting the *group's* existential well-being.[8]

As an example of the former, one can readily argue that the reason that communism in its abstract, philosophical form was as ideologically and socially seductive as it was—and it *was and still is* seductive—and thus much more readily propagated in the twentieth century than, for example, capitalism was or still is, is because it appeals to the basic human love of egalitarianism (see Boehm, 1999)[9] and fairness. It is so seductive that it is still adhered to by some groups,[10] despite the fact that a century of history has shown that it is outcompeted by alternative systems for organizing social groups larger than primary groups in intergroup economic competition. This is because it is ultimately rooted in indiscriminate generalized reciprocity (i.e., reciprocal altruism), which breaks down rapidly in groups as they become larger than a primary group due to a breakdown in reciprocity. This breakdown occurs because generalized reciprocity in particular is rooted in the value of the specific social relationships between the individual group members engaging in these exchanges and the social costs to those relationships that result from failures to meaningfully reciprocate. The basic problem is that those interpersonal relationships become more and more impersonal the larger the group and thus progressively less valuable. What do I, a resident of one urban neighborhood, necessarily care whether or not someone on the other side of town, much less another city or state, whom I will never meet face to face, and do not particularly need for any specific purpose, is angry at me for failures to reciprocate. If a member of my close kin, family, localized community, or one of my peer groups is angry at me for repeated failures to reciprocate, I have a problem because I value those interpersonal relationships for one emotional or practical reason or another; if some

[8]Which is not the same as saying it necessarily benefits the somatic well-being of all the group's constituent groups and/or individuals

[9]Just look at virtually every utopian philosophy promulgated in the west over the past several hundred years and you will see at least limited egalitarianism being a key element. That it was not successfully practiced in their attempted large-scale implementations is quite beside the point.

[10]Though not necessarily in the form of a soviet-style dictatorship of the proletariat.

stranger is angry at me, I do not usually have such a problem since I value those interpersonal relationships much less, *if at all.*

A good example of the latter is social structure and social organization, which are group-level traits that cannot possibly exist at the level of the individual, in that one cannot properly be one's own social peer, superior, inferior, parent, child, teacher, student, ruler, subject, neighbor, friend, fellow sodality member, fellow citizen, etc., nor can one person constitute a social group. Social structure and social organization affect the existential well-being of the group's membership—though not all members necessarily benefit to the same degree *if at all*—or some controlling subset thereof, in their competition with other groups (see Durham's, 1991 concept of imposition). One has only to look at the contributing factors in the relative success of European colonists at the expense of native American groups, the relative success of Niger Congo B (Bantu) speakers at the expense of indigenous Khoisan speakers in southern Africa, and the relative success of Proto-Indo-European speakers at the expense of the indigenous descendants of eastern Europe's Neolithic populations (see Anthony, 2007) to see evidence for such group-level advantages. Aside from technological advantage, *what all three instances have in common* is that the "winners" lived in larger groups, held together on a permanent basis by virtue of a more complex sociopolitical organization.[11] There is strength in numbers, and a more complex form of sociopolitical organization, aside from any practical organizational advantages that it itself confers, is a major contributing factor to having made those larger numbers possible.

However, it is important to keep in mind that despite the fact that the competitions that drive selective forces can be between groups as individual entities as well as between individuals, the sorts from all the hierarchies always flow through the nexus of the individual, as per Eldredge's (2002) sloshing bucket. That is, a reproductive or economic social group ceases to exist when the second to last member dies or abandons the group in favor of joining another group, because they have lost faith in the relative functionality of the group, and it thus ceases to any longer be a social group; a group's ideology ceases to exists when the second to last adherent jettisons it in favor of sharing an alternative ideology with a new set of others, because they no longer find it sufficiently satisfying, and it reverts to becoming just the single remaining individual's propriospect; and a group's skill pool ceases to exist when the second to last contributor to it ceases to practice their own skill set in the social or ideological context that had promoted it in favor practicing it in the context of another skill pool, because they have lost faith in the skill pool's relative efficacy, and what had until then been a group's skill pool reverts to just the last remaining individual's narrower skill set. The fact that selection can operate on groups does not alter the fact that the nexus of all these hierarchies is the individual or the individual's ideological or behavioral (i.e., "cultural") proxy. This is because the selective forces

[11]The depopulation of Native American groups due to imported diseases shifted the early size advantage away from them and toward the relatively small initial European colonial groups, but the importance of size advantage remains valid.

driving extinctions, be they extinctions of organic or metaorganic group-level entities, can only be directly experienced by the organic individuals that constitute them or their metaorganic proxies. They are only experienced by groups through the cumulative effect on the group's constituent entities.

Just to be clear, to say that selective forces can only be directly experienced by individuals is not the same as saying the resulting individual deaths or abandonments are of necessity a product of just individualistic action. It is clearly possible for the group to collectively decide to change its ideology, practices, social constitution, or even terminate its physical existence. Modern examples of a collective change in constitution include a decision to reorganize or dissolve a group through a corporate declaration of bankruptcy, and examples of group decisions to terminate their physical existence include the 1978 Jonestown mass suicide, the Solar Temple mass suicides in the 1990s, and the Balinese mass ritual suicides (*puputan*) in the first decade of the twentieth century, while the mass suicide of the last *Siqari'im* (aka *Sicarii*, Zealots) defending Masada against the Romans, and the mass suicide of the citizens of Xanthos defending their city against the Persians in 540 BCE are particularly famous older examples. However, while such decisions are/were made at the group level, it is/was directly experienced by the constituent individuals and only experienced at the group level through the group's decision's top-down cumulative effect on all its constituent individuals.

Also, it needs repeating that the competition between entities at any given group level is with reference to the entities as integrated, interdependent wholes visible to selective forces as entities of that level, just as the competition between individual organisms is between individuals and not their genes. Constituent lower order entities (e.g., individuals) do contribute to the relative success of the higher order entities of which they are constituent elements (e.g., groups) to one degree or another depending on the specific nature of the major higher order competition(s), and thus experience the top-down group effect to their relative success with respect to competing like-order entities (e.g., other individuals). But, at any given level, the competition is between like-order entities within a given hierarchy.

With reference specifically to culturally structured social systems, the raw material visible to selection is social *systems* of all scales: the more *functionally* complex a group's key social subsystems with respect to the major areas of competition with other groups, the more elegantly adapted its social system and the greater that group's advantage over other groups employing alternative social systems. On the other hand, the lower the level of embedded complexity within its subsystems (as typically in the difference between the donor and recipient of a diffused innovation) and the less constraining its Bauplan, the greater the system's potential for microevolutionary change in response to changing circumstances. Service (1960) in so many words made essentially the same points in his discussion of what he referred to as "evolutionary potential." However, to these should be added that the greater the level of embedded complexity, the more rigid the system and the greater the potential for stress-generated systemic failure and thus a punctuational evolutionary episode.

We need look no further than the first point to explain, for example, the spread of surplus-producing economic systems at the expense of immediate return systems. We need look no further than the second point to explain the differential adaptability and thus persistence of socioeconomic and sociopolitical systems that are essentially similar in terms of what are deemed (with respect to the issue at hand) to be key components (e.g., see Spencer, 1998). Finally, we need look no further that the last point to find the *ultimate* explanation for the long-studied (e.g., Tainter, 1988; Yoffee & Cowgill, 1988) collapse of state systems. State systems, by virtue of their very complexity, have a particularly limited capacity for incremental change as compared to simpler sociopolitical systems. If, due to stress (e.g., see Tainter, 1988), systemically unallowable change (e.g., the introduction of decentralized market forces into a highly centralized command economy, as happened in the former Soviet Union) is nevertheless persisted in, the system will collapse in whole or part. It will be replaced by a descendent system of lesser or equal embedded complexity, but having the potential (not necessarily ever realized by the descendent system) to develop a greater level of functional complexity. Unlike Herbert Spencer (1857), who saw rigidity as preventing change and ultimately leading to simply a culture's extinction (see Carneiro, 1973: 90), I see it as potentially precipitating bursts of rapid, largely unpredictable (in its specific sensu *lato* cultural outcome) change and the phoenix-like birth of new cultural systems along with the behavioral and social systems they regulate.

Chapter 7
Chance, Agency, and Loaded Dice

"History never repeats itself. Man always does." (Voltaire).

Chance and Evolution

Evolution, in the Darwinian instead of Spencerian sense of the term, is a historical process in that outcomes are not determined by prior conditions, even though they may be dependent on them. In other words, they are a product of chance. Chance, the mere possibility of an outcome as opposed to its certainty, is what distinguishes the operation of a truly historical process such as evolution from a progressively teleological one that is instead reliably predictable in its trajectory.

To begin with, evolution is historical because the various selective processes, the ultimate driving force in evolution, do not produce variants; they merely select from among the existing variants. Selective processes only screen what already exists; and, what exists for them to screen is only what has previously been produced by different processes. Just because a variant would be of selective value in a given environmental context does not mean it specifically will come into existence to be selected for, and simply put: if a variant does not already exist, it cannot be selected for by any selective process. However, the environmental forces at the heart of selective processes may also have an effect on what intent-driven behavioral variants come into existence due to the affordances they create and which such novel purposeful behaviors are then meant to engage with (see Gibson, 1979). To the degree such novel intent-driven behaviors have selective value, they would then lead to selection for any subsequent biological variants as may then come into being that facilitate any of those behaviors and thus actually also have selective value (see Walsh, 2015: 163ff.).

In biological evolution, a given environmental condition will not in and of itself trigger the novel appearance of any new *genetically* based variants, much less a specific variant that addresses that condition, let alone one that addresses it optimally. First of all, this is because the processes that produce new variants operate independently, though not necessarily *separately* from the processes that select from among the variants. As a result, an affordance can potentially be addressed in

M. Rosenberg, *The Dynamics of Cultural Evolution*, Studies in Human Ecology and Adaptation 12, https://doi.org/10.1007/978-3-031-04863-0_7

multiple if not myriad ways, some likely more efficacious than others with respect to any given environmental condition, but many theoretically having at least some degree of selective value regarding that condition, modified by whatever positive or negative value they have with respect to other more or less pressing environmental conditions. Selective processes will favor the most generally cost-effective variant available at any point in time and not wait for some optimal variant to appear.

Secondly, this is because the independently operating processes that produce the novel biological variants, on which selective processes can then potentially act, all incorporate some significant degree of chance in producing their outcomes. Moreover, the structure of the selective screen itself changes over time, typically also by chance. Finally, this is because both the processes that produce biological novelties and those that sort them are both sensitive to previous conditions. In other words, they are often enough contingent on previous outcome states.

Thus, evolutionary biological outcomes are always a product of chance at two concurrent levels: chance of current conditions being what they are and chance of some novelty coming into existence that better takes advantage of current conditions to produce some specific end than do other existing variants. The first is chance in the contextual, creation-of-current-potential affordances environment, as is integral to niche construction. Does the chance structure of the selective screen(s) at a given point in time favor some potential variants over other variants; whereas it might have favored different ones were current environmental conditions different because previous events had played out differently. The second is chance in the actualized ability to better engage with an affordance, in the sense of what actual variants—previously existing variants and any novelties—are available to be favored or not by the then-existing selective screen(s), and to what degree.

As noted above, the fact that a biological variant would have selective value does not guarantee it will exist to be selected for. The independent operation of the processes that produce novelties and those that select from among them sees to that. In the biological domain, what novelties exist for selective forces to operate on are, developmental processes aside, at least in part produced by mutation and other more or less random occurrences (e.g., crossing over). This, coupled with the randomness of contingent current conditions, makes the repetition of any given complexly contingent series virtually impossible, much less both the "contextual" and "actualized" contingent series running in a parallel repetition. As Gould (e.g., 1989b) was fond of pointing out in his popular writings with respect to the evolution of life on earth, were the clock somehow rolled back to some point in the distant past such as the "Cambrian explosion" of multicellular life ca. 542 million years ago and biological evolution allowed to play out all over again, the outcome would inevitably be radically different than the current one—of which our species is a prominent part. For example, in the case of our species evolution, some of those particularly crucial sequentially contingent chance events include (but are by no means restricted to) the chance absorption by some ancient prokaryote of the alpha-proteobacterium that would evolve into the endosymbiont we call the mitochondria of Eukaryotes; the chance shunting aside of the Synapsid ancestors of the mammals (the Cynodonts) in favor of the Sauropsid Archosaurs with the "great dying" extinction event ca. 252 million years ago that marked the end of the Permian; the chance shunting aside of the dinosaur descendants of the Sauropsid Archosaurs in the context of an asteroid impact ca. 63 million years ago that cleared the stage again for an eventual return to dominance of the Cynodonts' descendants in the form of the mammals; and the chance mutations that made language physically possible in hominins; not to mention innumerable other interspersed chance events, each contingent on an earlier outcome. Biological evolution is thus purely historical in its operation by virtue of current conditions' sensitivity to previous conditions and that kind of pure history never ever repeats itself in exactly the same manner. If the clock were turned back and allowed to run again, we would very likely not be here to inquire as to what happened when it struck a particular hour much less to inquire as to how we came to be.

Cultural evolution, as an evolutionary process, is also historically contingent in its outcomes and thus subject to chance by virtue of sensitivity to previous conditions, including those that shaped a given system's Bauplan. Minus the explicitly evolutionary framework, this historical view of culture change goes back to Boas and his students. However, chance is simply the absence of certainty, meaning that it need not be of necessity completely random. In fact, a good case can be made that the role of random chance is often not as great in cultural evolution as it is in biological evolution. More specifically, a good case can be made that the production of behavioral novelties (for selection to act on) may not be as random as it is in biological evolution, in that the production of behavioral novelties is often affected by behavioral biases influencing purposeful actions to produce a high degree of parallel evolution in the cultural domain. Witness, for example, the invention of agriculture in multiple independent locales at the end of the Pleistocene and the apparent independent evolution of pristine urban states (contingent on the earlier invention of agriculture) in at least three parts of the ancient world: the Americas, southwestern Asia, and eastern Asia. This is because, for species with a degree of sentience, agency is integral to behaviorally engaging with affordances and choice of actions is influenced by evolved behavioral propensities.

A view of cultural evolution that relies only on selective processes and accords no processual role to purposive novel behaviors implicitly requires a human population to passively maintain an existing behavioral complex in the face of mounting stresses as the contribution to biological fitness of that behavioral complex decays. This must continue until such time as it and some novel or hitherto less adaptive alternative complex are made equally viable by deteriorating conditions, at which point the alternative complex will start being selected for. In the Modern Synthesis, such environmental stresses were generally thought to have no significant impact on the appearance of novelties because organisms have no control over their biological processes (but e.g., see Walsh, 2015). This is not so in the case of cultural evolution because of its Lamarckian elements, which derive from human agency in the context of human cognition. In a sentient species such as humans, stress both generates a variety of innovative behaviors and independently selects from among them. Thus, the rate of innovation will vary continuously and directly with the level of stress and some innovations, by the very nature of selective processes, will be favored by selection. In other words, innovation will usually begin long before the behavioral system has become anywhere near severely dysfunctional and a case can be made that such stress-induced innovation will often be biased in the direction of addressing the perceived manifestation of those stresses.

As noted in the previous chapter, most evolutionary and transformational approaches to cultural evolution have at their core the implicit assumption that a cultural system is completely open-ended in its potential for change. For some, this typically goes hand in hand with the implicit assumption that all potential novel behaviors have an equal chance of occurring—in the specific case of evolutionary approaches, to then be selected for or against. Leaving aside transformational approaches for being progressively teleological, the question, however, is: do all potential novel behaviors actually have an equal chance of occurring? That is, leaving the Bauplan's constraints aside for affecting only cultural selection, do all possible behavioral elements have an equal likelihood of appearing to become available to selective forces? Are novel behaviors as "purely" random in their appearance as are novelties said to be in the biological domain; or, are there proximate processes at work that tend to bias the chance appearance of some behavioral novelties over others, making them more likely than others to become available to selective forces whether or not they are subsequently permitted by the system's Bauplan. In other words, are the dice that dictate actualized chance as regards novel behavior in some way "loaded?" And if they are, that raises the secondary questions of determining how, in what manner, and to what degree the dice are so loaded.

Chance and Causation

Evolutionary processes are characterized by the operation of chance and thus not reliably predictable, while transformational progressive models are characterized by consistent, if not nomothetic causation, and are therefore reliably predictable. Evolutionary models in general and more specifically those dealing with the mechanics of cultural evolution, being truly historical, have at their heart the premise that causation is uncertain; outcomes are complexly determined through the infinitely variable interplay of contextual chance, the chance that environmental conditions will favor any given novelty as may appear, and actualized chance, the chance a given novelty will appear and thus become visible to selective forces. On the other hand, transformational models of cultural evolution have at their heart the premise that, for any given type of phenomenon, causation is consistent and thus clearly definable, as well as widely applicable, involving at most only contextual elements of chance—is the opportunity there or not for some behavioral novelty to be engaged in, as it inevitably will if the opportunity exists. Thus, for transformational models, a straightforward and direct cause is theoretically discoverable for any given repeated phenomenon. The only epistemological uncertainty revolves around whether any given proposed contextual cause for some such phenomenon can be accepted as *the* cause.

What distinguishes an evolutionary process in the biological mold from a transformational one is that the former requires that the chance-influenced forces that make something manifest (its proximate cause) operate independently of the chance-influenced forces that favored it over the other available varieties (its ultimate cause). As most influentially laid out by Mayr (1961), for biological traits including those that underlie innate behaviors, the cause for the expression of some trait is its proximate cause and distinct from its adaptive function—the cause for its selection, which is the ultimate cause for its evolution (see also Birch, 2017: 4ff; Uller & Laland, 2019). In other words, as encapsulated by Uller and Laland (2019: 2), for Biology (in Mayr's view), evolutionary biologists are concerned with ultimate causation while molecular and developmental biologists as well as most other biologists are concerned with proximate causation.

To say that some species of birds migrate seasonally to (among other things) maximize access to food, while simultaneously also minimizing predation on their young is to present the ultimate explanation for the phenomenon—its function; it is an evolved adaptation produced by selection for what it functions to accomplish—relative reproductive success. Birds that did it were more successful at reproducing (and passing on the propensity for that behavior) than those that did not. To say that the behavior is innate and based in some evolved physiological, developmental, or cognitive process that triggers such migration is to explain the behavioral propensity's proximate cause. But, that is still subtly different than asking what environmental factors actually trigger the innate behaviors (e.g., temperature, daylight, hunger, the young have fledged, etc.) in a particular individual or group of birds at a particular time in a particular place—the actual behavior's specific expression in time and place. That is asking about the proximate cause's trigger, as opposed to the propensity's proximate cause in, for example, developmental processes, etc.

To the degree that a given species lacks cognitive abilities, the trigger for the behavior must in and of itself be the environmental stimulus. The conditions produce the behavior—if A, then B. But, to the degree a given species has cognitive abilities, the stimulating conditions and resulting behaviors must be mediated by cognitively filtered intent, and thus the resulting behaviors are potentially variable. Thus, broadly speaking, proximate causation in a sentient species—and thus humans most of all—actually encompasses several related things: the biological basis for a behavior, its specific environmental trigger(s), and the agents' intent. All are integral to any discussion of cultural evolution, wherein motivation underlies agential actions and can potentially produce contra-cultural behaviors that precipitate the birth of new cultural systems. Were it not for agency, we would still be a population of forest-dwelling apes like our late Miocene distant ancestors were, eying potential resources out in more distant, more open, and more alien/dangerous country but unable or unwilling to try their hand at tentatively exploiting them in one manner or another. After all, there had to be an individual who first took the small step that was the giant leap.

However, for cultural evolution, causation is a bit more complex than even that. This is because ultimate causation can be said to be situated in the economic hierarchy as well as *both* of the two "reproductive" hierarchies: the organic genealogical hierarchy governed by sexual selection and the metaorganic information hierarchy governed by cultural selection. In the organic hierarchies, we find the adapted mind and particularly its behavioral propensities (e.g., see Cronk, 1999; Richerson & Boyd, 1999), while in the information hierarchy we find the prescribed and proscribed nonbiologically based behavioral constraints, all produced by selective forces.

Proximate causation is the motivation to behave in a particular way, novel or not, influenced by the interaction of those propensities and constraints. The outcomes of those behaviors, in the form of a given behavioral variant's relative real and/or perceived[1] functionality flow back to the "reproductive" hierarchies via the BSs → CS and BSs → SS sorts (in the case of the latter, sometimes as BSs → NS/NSs → SS) where they affect evolution as ultimate causation.

The motivation to act can be rooted in the sorts flowing to the adaptive hierarchy from any or all of the other three hierarchies; selection for the persistence of or resistance to the behavior can be in any or all of those three as well. The point here is that proximate causation for behavior in a sentient species and humans in particular is neither simple nor necessarily singular. For behaviors, proximate causation has the potential to be complex and sometimes contingently sequential, in that it can involve innate behavioral tendencies, selected for on the basis of their inherent math, which affect calculations both in that initial context as well as any subsequent chain of contingent contexts.

Lastly, bypassing Walsh's (2015) assertion that proximate and ultimate causation can interact biologically as interesting and certainly relevant, but unnecessary to explore for present purposes, it should be noted that to say that proximate and ultimate causality must operate independently of each other is not the same as saying that the forces that drive them must necessarily be *different* from each other. While those forces are in fact both independent *and* different (if not truly separate, as per Walsh) in biological evolution, that is not necessarily the case with cultural evolution. All that is truly necessary is that proximate and ultimate causation be acting independently of each other, with environmental pressures operating on entities in the organic hierarchies working *differently* to, on the one hand, stimulate purposeful behaviors that attempt to address those pressures (with no guarantee that they do so successfully much less optimally), and on the other, selecting from among novel and existing behaviors with respect to how well they address those pressures. To appropriate a figure of speech: all that is really necessary is that the right hand not know what the left hand is doing and vice versa, whether or not both hands are parts of the same body or not (as in the fractionated view of the Modern Synthesis).

---

[1] Perceived functionality affects cultural selection, which can potentially be at odds with sexual or natural selection, which are based on real functionality (once again, refer to the Shaker example).

In short, proximate causation is the effect of the sorts from the organic hierarchies coming back into the organic hierarchies via the metaorganic hierarchies, having been refined by them to be the sorts from those hierarchies. Thus, as regards behavior, it is entirely consistent with a proper understanding of evolutionary processes to propose that environmental[2] stimuli can both *tend to* trigger behaviors, including behavioral novelties—all subject to being influenced by innate behavioral propensities, and at the same time those same environmental forces can independently select from among any novelties so produced, as well as from among any also produced randomly.

Proximate Causation and Loaded Dice

Within evolutionary approaches to cultural evolution, there are two basic approaches to the way chance operates with respect to proximate causation and the production of novel culturally patterned behaviors. One, by default, implicitly holds chance to be equal for the appearance of all potential behavioral novelties which then become available to selective forces; the other, favored here, explicitly holds chance to be biased toward some potential novelties appearing over others. The first is most often integral to studies that restrict themselves to the archaeological record and within that framework ones that focus largely on cultural reducta in the form of individual traits (or the behaviors that produce them) and the specific workings of the organic hierarchies on such reducta. The second is integral to studies that focus on the systemic workings of cultural evolution and acknowledge Lamarckian elements as inherently having a role in cultural evolution by virtue of (in this case human) sentience.

The first is more often than not used for explaining the artifacts that constitute the direct archaeological record itself, with sole reference to ultimate causation—the history of what was selected for and why it may have been so selected for. Proximate causation—how something came to exist for selection to act on in any given case—tends not to be addressed. The rationale for that position is that it is only the "phenotypic" (i.e., phenomenological) manifestations of culture (i.e., behavior), archaeologically expressed in the form of artifactual residues, that are/were visible to selective forces in the organic hierarchies. Natural and sexual selection can only act on what they can "see" and it is held that they cannot directly "see" the information in the information hierarchy, much less how the CSs → BS sort from that hierarchy is filtered by entities in the adaptive hierarchy, leading to motivations to behave in a given manner. Moreover, from a reductionist perspective, the information systems themselves (i.e., cultural entities) are considered to be on the one hand merely abstracted ephemera constantly in a state of becoming and never actually existing, and on the other hand, even if they were proven not to be

[2] Which includes the social environment

ephemeral, they are not recoverable archaeologically in the sense of not being visible to archaeologists.

Thus, agency is rejected as a legitimate element in the explanation for behavioral novelties on the grounds that attributing intent is not falsifiable archaeologically and thus not knowable scientifically (e.g., Dunnell, 1989; O'Brien & Lyman, 2000). All novelties are thus not so much said to be, but by logical consequence necessarily treated as having an equal chance to arise in actuality. After all, the variant in question had to have arisen to be selected for, as evidenced on both contextual and actualized counts by its presence in the archaeological record, and that is considered sufficient for the kind of paleontological-style historical explanation being sought.

While the internal conditional logic integral to this view is generally sound so far as it goes, it is arguably built on three faulty assumptions that unnecessarily restrict its purview. The first assumption regards the mechanics of evolution and it stems from this view's reductionist tendencies. Just as reductionism in biological evolution is required to proceed by individual gene replacement, here it is required to proceed by individual trait replacement. Thus, this view only addresses what is here being considered microevolution and makes no attempt to deal with what is here considered macroevolution, which often requires both intentional deviation from critical norms that would not normally be attempted and consequently *systemic* replacement.

The second assumption is that selection in the organic hierarchies cannot "see" the sorts from the information hierarchy, except as their materialistic manifestations in the form of behaviors engaged in by organic entities and visible archaeologically as artifactual residues. It needs to be noted here that this is a statement about natural and sexual selection's ability to "see" information, not the archaeologists' ability to "see" information.[3] The flaw here is that it does not take into account the sorts flowing from the information hierarchy directly to selective forces in the two organic hierarchies (CSs → SS and CSs → NS). Thus, both sexual and natural selection *do* have the ability to "see" and select from among competing informational entities without the sorts from the information hierarchy first being mediated by selective forces in the adaptive hierarchy.

The third faulty assumption is that: because information cannot be "seen" archaeologically, intent is not knowable scientifically. Even if we were to allow the highly debatable point that information systems cannot be seen archaeologically (see last section of chap. 4), and that therefore proximate causation in the form of intent is not knowable *archaeologically*, that is not the same as claiming that it is not knowable *scientifically* – there are, after all other sciences besides archaeology. In other words, this assumption conflates archaeology with science in general; it does not allow that the data provided by other scientific disciplines, such as Psychology and Cognitive Ethology—particularly primate ethology can demonstrate general intent and aggregate agency (e.g., De Waal, 1982, 1996; Kohler & Gumerman, 2000).

[3]The archaeologist's presumed inability to see information systems is also highly debatable (see the last section of chap. 4), but that issue is both too tangential and too unwieldly to address adequately in the current framework.

True, what a given individual specifically intends by a specific action is not reliably knowable either archaeologically or scientifically unless communicated verbally[4]—a given individual's mind is a "black box," the interior of which cannot be reliably viewed by an observer. Specific intent can vary depending, for example, on whether or not that morning the individual argued with someone, stubbed their toe, or experienced what they perceived as an omen. For want of a nail the shoe was lost, for want of the shoe the horse was lost, and so on.

Specific historical outcomes are not predictable because an individual's specific intent is not knowable. However, aggregate general intent is another matter entirely because it is at aggregate levels that behavioral tendencies, as elucidated by other disciplines, are *recognizably* expressed, including archaeologically, and humans are if nothing else social animals. Culture change is a social process. On person acting innovatively is simply one person acting in a deviant manner, as was Rakudu, and if that behavior is not adopted by others then culture change does not take place. Culture evolution does not result from any given individual's agency, but from aggregate agency.

Behavioral *tendencies* are observable in the aggregate and their typical purposes are inducible from aggregate outcomes. Thus, they are at a minimum scientifically knowable, and therefore applicable to archaeological settings even when not directly observable archaeologically. Black holes, the Higgs field, or bottom quarks are all not *directly* observable by astronomers, astrophysicists, or particle physicists either, but they are nevertheless indirectly knowable by their effects. In short, behavioral biases can be observed in humans or even in other species by their effects. And, to the degree those observable in other species can be reasonably extrapolated to human behavior, all can be applied to archaeological contexts that reflect aggregate behavior—in other words, most archaeological contexts.

Lastly, behavioral tendencies bias the chance a behavior will appear to be selected for or against; they load the dice in favor of behaviors being employed for which there is a bias. Even so, proximate causation is not precisely predictable because behavioral tendencies only *favor,* they do not determine the choice of certain behaviors over others when the physical and/or social environmental contexts potentially trigger them *and* make them seem to be viable behavioral strategies for agential actions. Given the biases favoring their employment, they are highly likely to eventually be triggered, but only *if nothing tried beforehand* is seen to effectively address the forces reflected in the sorts driving behavioral selection to a significant enough degree. In other words, while behaviors rooted in behavioral tendencies are more likely than other behaviors to be tried first, there is no guarantee they will be the first employed, though they are likely to eventually be tried unless circumstances specifically prevent this or some other behavior tried beforehand is perceived to work well enough for now.

[4]This raises the secondary question as to whether you can reliably take the individual at their word due to the possibility of self-delusion or outright lying, but that also is a tangential issue.

Proximate Causation and Contingency

Proximate causation is sensitive to previous conditions, which are not necessarily the immediately previous ones and which may have been produced by the same or other behavioral tendencies. A behavioral choice made in response to a set of current conditions will at once set the stage for and constrain choices made subsequently in response to future conditions contingent on those current conditions; a different choice would have both created a different set of future conditions and differently constrained future choices to such future conditions. Such future choices are also subject to behavioral biases, be they the same biases as influenced previous choices or different ones.

While behavioral selection is biased by innate tendencies, those biases will not necessarily result in identical choices, even by groups presented with an essentially similar range of affordances. The variability in specific local conditions and the fact that we are talking about biases and not determinants, coupled with the ultimately imprecise ability of individuals to calculate costs-benefits and the constraints imposed by both Bauplan and previous choices, all see to that variability in choices. So do the chance elements of what is produced in punctuational episodes. Behavioral selection is thus at once a branching process, one capable of producing divergent outcomes, as well as a channeling process capable of *eventually* producing somewhat convergent and parallel outcome states and thus a degree of *functional*, though not necessarily structural equifinality, due to biased behavioral tendencies operating to eventually produce somewhat similar sequentially contingent outcomes and the resultant patterns in prehistory.

Just to be clear, functional equifinality due to similarity in ultimate outcomes does not imply unilinearity. This is because contingent sequences potentially play out variably intermixed with and interrupted by other contingent sequences, some potentially also affecting what choice nodes then become available in other sequences and others not. It just means that a given (often biased) behavioral choice is contingent on a previous (often biased) behavioral choice, whether or not the latter was immediately previous. That relevant previous choice could have been made at any previous point in a long sequence of choices, not all contingent on the same previous choice, or even contingent on any previous choice at all, and not necessarily having been made immediately previous to the affordance whose availability it is contingent on it. That is, while one group may theoretically have made the (often biased) set of choices in uninterrupted sequence, another may not have made the critical (for convergence) contingent choice until much later in their sequence of choices, having made other (often differently biased) choices—some independently also contingently sequential, but again not unisequential—in the interim. This would have put them on a divergent trajectory until such time as the contingent choice for potential convergence is also made by them and the degree of convergence in outcome states becomes *potentially* possible. Still other groups may never make the critical-for-convergence contingent choice and continue on the divergent trajectory. It is the contingent sequentiality that produced the patterns in prehistory and it is the lack of unisequentiality that negates the idea of cultural evolutionary stages. Instead, what we have are just multiple contingent threshold states of being upon which in one combination or another other potential states are contingent, but which interim states of being can come into existence at different earlier or later times because they are not all part of just one single overarching contingent sequence.

Human Nature and Proximate Causation

Born in part as a reaction to the Social Darwinism of the late nineteenth through early twentieth centuries and the broader racialist/racist theories that held sway in the west from at least the time of Linnaeus and his racial taxa until the middle of the twentieth century, feeding slavery and then later segregation in the United States and culminating in the Holocaust in Europe, the view that there is no appreciable thing as human nature was briefly popular in the social sciences during the first few post-WWII decades. In this view, we, as individuals, are more or less *tabula rasa*—a product of nurture, of which culture plays the single most prominent part, with no significant input from nature. It has long since been largely discredited (e.g., see Degler, 1991),[5] though the effects of that view linger on (see Cronk, 1999). Human beings do have certain innate mental and behavioral tendencies, some of which have been studied under the umbrella of Evolutionary Psychology and attributed to the adapted mind, others under the umbrellas of Ethology and Human Behavioral Ecology.

To recapitulate, Evolutionary Psychology's concept of the adapted mind holds that the human mind is itself a *biological* adaptation shaped by natural selection over the *full* course of human biological evolution and that this adaptation is phenotypically expressed as behavioral (and cognitive) propensities. As an adaptation not of necessity applicably restricted to the specific foci of Evolutionary Psychology, any other innate mental or behavior tendencies discerned by other disciplines can logically also be attributed to similar adaptations, if not simply to other aspects of the adapted mind. Thus, the corner stone of agency—the capacity for (consciously or unconsciously) calculated decision-making can be considered such an adaptation. These adaptations imply that there exist both innate and calculation-generated behavioral tendencies biasing the actualized chance that a given kind of behavior

[5]For example, the idea that, if one did not give boys toy weapons to play with, they would not engage in simulated play violence (and thus be less prone to engage in real violence as adults) has been shown to be wrong; they simply used a hand with extended forefinger to simulate a gun or a piece of wood to simulate a sword. This is simply to point out a specific behavioral tendency and perhaps one more so expressed in males than females (e.g., see Wrangham & Peterson, 1996). This is not contradicted by another fact: that the expression of innate tendencies can be checked or made more pronounced to at least some degree by cultural expectations. Culture-bearing individuals are not prisoners of their genes as regards behavior. Both nature and nurture have their roles in shaping it.

will appear in a given situation and hence become available to the operation of selective forces. In other words, such tendencies tend to bias behavioral selection and thus proximate causation.

One such tendency is the tendency to actively compete in the face of environmental pressures, which environments typically include conspecifics.[6] In the case of humans, others include the tendency to cooperate, to act altruistically (often a form of indirect self-interest), to act self-interestedly when not acting altruistically, and to promote peace/harmony within the group (by force if necessary). All these tendencies have been selected for in species wherein they constitute some part of the behavioral repertoire because they confer some form of advantage to the species' constituent members either individually or top-down via membership in the collective. None of these tendencies are uniquely human and some such as the tendency to actively compete are ubiquitous in the animal kingdom.

Do not conflate the Social Darwinist view of competition—the striving to progress of races/nation states/classes in a zero-sum biological game—with a proper evolutionary one that appropriately accommodates culture. That is, in the former view, the characteristics of the entities so competing were viewed as biologically fixed with respect to all physical *and behavioral* characteristics. In the latter view, we are also dealing with the competition between malleable and permeable informational entities and the behaviors they tend to produce and which behaviors vastly outnumber the behaviors that are biologically generated. In the former, competition produces progress because superior groups will displace inferior groups; in the latter it merely produces informational entities that propagate more readily and can be readily adopted by any biological group that now sees it, rightly or wrongly, as superior to what it replaces, often based on the informational entities' impact on somatic well-being.

Sentient animals, the more sentient the more so and thus humans most of all, when confronted with environmental stimuli—and competitors are very much a stimulus to act in that they stand in the way of achieving a given desire—attempt to address the stimuli at either the individual or group level. If it can be addressed adequately through existing behavioral norms, it will most likely be so addressed. If it cannot, then individuals may be motivated to innovate and such innovations will be intended to address the competitive problem *as it is perceived through the lens of the information system constraining behavior*. Moreover, many species, and humans most of all, have the ability to consciously or unconsciously calculate some variably accurate approximation of the likely costs and benefits of a given behavior. Such ability to calculate is a cornerstone of Cognitive Ethology and Behavioral Ecology (including Human Behavioral Ecology) and it means that the behavioral novelties produced in a given situation will be biased in favor of ones that may actually have

[6] Just to be clear: this should not be taken as a stereotypical statement (i.e., all X are/do Y). It is a statement of tendency—of bias when confronted with a choice of potential actions. Any given tendency is expressed variably even in individuals of the same species because they are, after all, individuals and thus in part also unique phenotypical products of their individual histories (i.e., nurture).

some level of benefit relative to the social and/or environmental stimuli that triggered them.

The key distinction between such a view of intent-driven innovation and the intent inherent in earlier progressive and adaptationist transformational views of cultural evolution is that the intent here is neither to progress toward some goal nor to adapt in some foreknown-to-be-optimal manner. The intent is merely to respond in calculated fashion to the perceived manifestations of environmental stimuli. Those calculated responses are not *invariably* optimal, nor invariably even selectively valuable; they are simply calculated responses to environmental stimuli that take into account both social and material costs and benefits. How selectively valuable such responses are depends heavily on both a given entity's ability to competently calculate and the information system's filter through which those stimuli are both perceived and understood, not to mention by which the relative value of potential outcomes are weighed. However, such responses, including innovative responses, will tend to be biased by the species' behavioral tendencies, which to some degree favor some behavioral responses over others and some innovative directions over others, whether or not they fit within the constraints of Bauplan. Such tendency-biased responses would tend to have a greater chance of actually having selective value in at least some contexts by virtue of the simple fact that *innate* tendencies are just that because they are rooted in biological adaptations fixed over time by the selective forces in the organic hierarchies. If they were not a given species' biological adaptations, they would not be *innate* tendencies. But, it is important to remember that tendency-biased responses do not invariably have selective value in a given context, since biased responses certainly have the potential to be inappropriate for the situation at hand, to the point of potentially being completely counterproductive (i.e., maladaptive in that context).

The point is that, while there is an element of chance at work here as well, the actualized chance of a particular broad type of novelty being available to the operation of selection is biased, not equal. Equally important, so is the chance that the novelty produced by such biases would have some degree of selective value in the context of the situation that precipitated the innovative behavior. The degree of such bias in favor of a novel behavior having selective value is variable, context-dependent, and related to the clarity with which *apparent* cause and effect can be objectively linked to the situation consistently, with a higher degree of clarity producing a higher chance of bias in favor of a novelty actually having enhanced value. That is, for a problem involving hunger, cause, and effect (and thus the choice of potentially efficacious responses) are typically understood with a higher degree of clarity than would be normally the case for a problem, such as disease, interpretable using a wider variety of theories. For the former, the typically obvious solution is to somehow get sufficient food (with biases affecting the choice of how to go about doing so); for the latter, in the absence of some kind of "germ theory," a novel response involving supernatural powers is perhaps just as likely to be employed as is a response involving a new kind of herbal remedy. Moreover, the degree of bias in favor of enhanced selective value is increased by the degree to which stimuli are cognitively experienced in ways that facilitate an *objectively accurate* conceptual linking of cause and effect (again, hunger versus disease). That is, the more that the relevant understandings of cause and effect are objectively correct, the greater the chance that the triggered biased responses would be appropriate to the situation and be patterned in ways that produce perceptible benefits to the individuals practicing them and thus the responses' likely perpetuation via the various modes of selection.

Competition and Evolution

Robert Malthus's (1798) treatise on population and economics (and their interplay's implications for the possibility of social progress) was hugely influential in the formulation of Darwin's theory of evolution driven by selective forces and differential reproduction. In a nutshell, Malthus proposed that there is a natural tendency for populations to grow until checked in such continuing growth by increasing resource scarcity and such scarcity's consequences for somatic well-being.

> The power of population is so superior to the power of the earth to produce subsistence for man, that premature death must in some shape or other visit the human race. The vices of mankind are active and able ministers of depopulation. They are the precursors in the great army of destruction, and often finish the dreadful work themselves. But should they fail in this war of extermination, sickly seasons, epidemics, pestilence, and plague advance in terrific array, and sweep off their thousands and tens of thousands. Should success be still incomplete, gigantic inevitable famine stalks in the rear, and with one mighty blow levels the population with the food of the world. (Chapter VII)

Moreover, Malthus argued that increases in resource availability will just result in further population growth made actualized by said increased resources, rather than meaningful economic improvement over what had to the point of such increased resource availability been a smaller population held in check at that level by the unavailability of additional resources. In other words, population will inevitably tend to grow to utilize all available resources to the point of impactful scarcity, in the same manner as (in the pre-digital era) an academic's collection of books and journals would typically grow to fill the available shelf space (and ultimately all but the most essential flat surfaces). Adding shelf space would have cleared the flat surfaces for a time until they were once again covered by new additions to the collection for which there was now once again not enough shelf space. Such population growth is not necessarily required to be steady and can even theoretically be punctuated by periodic declines not directly triggered by Malthusian penalties, but the trend over time is toward growth.

Such perpetual latent, if not manifest, scarcity inherently creates competition, and in that sense it would probably not be an overstatement to say that Malthus' insights (extrapolated out to all other species) were the single most essential element in Darwin's formulations. This is because without competition, selection would be largely ineffective, and without effective selection there would be no evolutionary force in operation beyond drift and thus no such thing as adaptations produced by selection. That is, without competition all variants could then maximize their inherent functional potential without an actualized loss of some part of that potential to more efficient or otherwise better adapted variants. The result would be that forces favoring those superior variants would be significantly weaker at best and nonexistent at worst. If one believes that selective forces drive biological evolution, one is required to believe that competition, though capable of varying widely in intensity, is central to any proper evolutionary process, largely constant, and ongoing. To deny the ubiquity of competition is to deny the mechanics of Darwinian evolution, whether applied to the organic or metaorganic domains.

All organisms compete for resources and those that reproduce sexually also compete for mates. Such competition can range on a sliding scale from merely latent or barely perceptible to violent. Social species also compete for social standing because it secondarily relates to both somatic well-being and reproductive success. That competition drives evolution in the organic hierarchies in the form of natural and sexual selection. Social animals must balance that competition with the variable degree of cooperation that constitutes the competitive advantage conferred by that species' degree of sociality on either the group's constituent individuals directly or via the top-down benefit of their membership in the group as a whole.

Just to be clear, *competition does not disallow cooperation* and vice versa; cooperation can and usually does coexist with competition. For example, family members are latent competitors for resources within the family and latent cooperators in their family's competition with other families, not to mention latent competitors with some family members while cooperating with other family members in that intrafamilial competition. Similarly, individual families are latent competitors within their community, as well as cooperators with other families in that intracommunity competition, and latent cooperators in their community's competition with other communities. The same potentially applies to any kind of entity at any scale.

Also, again just to be clear, in human societies cultural norms can dampen or redirect, if not successfully suppress that competition (e.g., see Boehm, 1999). They cannot, however, suppress the innate *tendency* to compete; if they could successfully suppress the underlying tendency, there would be no need for the norms that attempt to suppress the competition itself—one does not go to the bother to build bear traps if bears are not a problem. The power of that tendency to compete is clearly visible in the costly displays of status we often engage in even when there are highly unlikely to be any tangible benefits to such displays because the target of the displays is the world at large rather than specific individuals with whom we are or wish to be in either a cooperative or *meaningfully*[7] competitive relationship.

Yet, Carneiro's (1970) proposal regarding state formation notwithstanding, it is only within approximately the last quarter century that human competition has begun to get significant attention as regards its role in cultural evolution (e.g., Wills, 1988, or more recently e.g., Field, 2004). One could argue that this is in

[7] Meaningful is here defined as being a competition having the potential to actually yield perceptible benefits.

part because, at least for evolutionists, with their primary focus on ultimate causation, it is of little obvious interest. Competition is implicit in the mechanics of evolution and thus why it evolved does not require explanation, unlike say the evolution of cooperation (e.g., Axelrod, 1984; Bowles & Gintis, 2011), which until the late twentieth century seemed incapable of being fixed as a group-level biologically based characteristic by the mechanics of biological evolution and therefore cried out for explanation. However, it is just as likely that in some quarters the reluctance to deal with competition's, and particularly violent competition's (see Keeley, 1996; LeBlanc, 2003) role in cultural evolution, was and perhaps still is its lingering evocation of the early twentieth century militarism, justified by Social Darwinism, and which led to the carnage of WWI and WWII. Alternately, in other quarters it may be simply philosophical incompatibility, because it gives nature major standing alongside nurture as regards human behavior (e.g., see Pinker, 2002; Wrangham, 2018), in this case as regards a particularly distasteful behavior for many. That is, an acceptance of an innate tendency to compete, violently if necessary, constitutes an obstacle to the belief that humans are perfectible via nurture (e.g., see Adams, 1998). This is not to denigrate that belief; behavioral tendencies may have been acted on in the past to produce the present and they may very well influence the future, but they do not determine that future; history is, after all, historical. It is simply to point out such beliefs as a possible subtle cultural influence that perhaps explains some of that reluctance to address the ubiquity of competition and particularly violent competition over resources in the past. The same people who have no difficulty in accepting that the recent Gulf wars were fought over who controlled the region's oil often enough have much greater difficulty extrapolating such violent competition over resources to simpler societies, and the simpler (and more noble the "savage") the more so.

While ethnographically or historically documented instances of violence are often enough readily acknowledged for simpler societies, the debate tends to be framed in terms of evidence for *warfare* (e.g., Nakao et al., 2016) and skeptics cite the lack of sufficient evidence to conclude that warfare occurred often enough for instances to not be considered anomalous. But, framing the debate in those terms sets up something of a straw man. Warfare is *organized* violence and thus only to be regularly expected among sufficiently large and sufficiently organized groups and even then only under sufficiently strong competitive pressures, which cannot be mitigated by other, less risky means (see chap. 8). Violent competition is inherently very risky and potentially very costly. Thus, it is to be avoided when possible, ideally begun at the lowest appropriately possible level of (non)lethality, and ideally ended as quickly as possible before it can become deadly and/or escalate into a protracted blood feud or something on an even more serious social scale that jeopardizes social networks (e.g., see Lee, 1979; Boehm, 1999).

The point is that up until the more recent prehistoric past groups were usually neither large nor organized enough to conduct what we would consider "warfare" and, more importantly, the threat of escalation was more often than not sufficient to achieve the desired effect of having one of the competitors—whether an individual or a group—quit the confrontation at least temporarily before more people were actively drawn into it. This is because in acephalous societies, the more people involved in a violent conflict of interests the more difficult it becomes to end and ending lethal competition as quickly as possible is usually highly desirable, at least for the weaker party in the dispute. The absence of warfare or even the absence of large-scale individual violence does not mean that violent competition did not occur—there is sufficient evidence of that. It just means that more often than not it was likely confined to nonlethal means if possible (e.g., threat displays), for good reason, and if violent, as restricted in scope as possible.

Competition Versus Population Pressure

For animals and more specifically humans, competition manifests itself most directly in three fundamental areas of life: subsistence, reproduction, and for social animals also social standing, which affects both of the first two, with biases in the appearance of behavioral novelties being most pronounced toward those ends of the behavioral spectrum relating to these fundamental areas. This is because, on the one hand, the centrality to life of these areas makes it likely that there already exist the greatest number of behavioral tendencies pertaining to them (i.e., the adapted mind) available for exaptation. On the other hand, it is because these domains are most central to most people's lives and thus subject to the most attentive (which is not to imply necessarily the most accurate) calculations of cost-benefit.

Competition is a product of the limited availability of something desirable relative to how many desire it. It is simply the result of an imbalance between the supply of something desirable, be it resources, standing, or best/most potential mates, and the demand for that something desirable—more people than can be satisfied with the existing supply of that something. When that imbalance is in regard to resources and the somatic well-being that those resources provide, it is nothing more and nothing less than what was called population pressure, however naively that term was employed as a "prime mover" in the context of the first generation of causal models for the evolution of food production that were proposed in the 1960s and 1970s as a replacement for the progressive discovery models that preceded them.

Critics of population pressure as regards its only specific application as self-contained explanation (i.e., prime mover) for the origins of food production, aside from rightly criticizing the evolutionary, ecological, or social mechanics of these early models, also tended to focus on two evidentiary lines, one archaeological and the other demographic. The first was the claimed lack of archaeological evidence for it either in the form of violence or in the form of population growth leading up to food production's inception. The evidence for violence will be addressed momentarily, but, as astutely noted by Bettinger (1998), population growth should not to be taken as a measure of population pressure. In fact, given the tendency of populations to grow, the absence of population growth does not indicate the *absence* of population pressure. It indicates quite the opposite, in that it is the clearest measure of population pressure, because it indicates the operation of Mathusian checks on further growth; the slower the growth the greater the population pressure, with no growth or even negative growth indicating extreme population pressure in a given environmental context. The same inversion of the criticisms somewhat applies to evidence of violence. Yes, the presence of violence is obviously evidence of intense competition (i.e., population pressure), but given the costs/risks associated with it, we should only expect it when less costly/risky alternatives (e.g., fissioning, intensified resource extraction, threat displays) are either unworkable or already at their limits. The situation for the terminal Pleistocene hunter-gatherers of the Nile Valley with its associated high levels of violence is a particularly good case in point of a subsistence system at its productive limits, with no obviously viable way to either shed excess people or innovatively diversify or intensify production until the introduction of domesticates several millennia later.

The demographic argument that critics presented was the claimed ability for human groups to regulate births and thus obviate population pressure. Even if we grant the dubious proposition that this could voluntarily be done effectively over protracted periods of time in the face of a likely tragedy of the commons due to defections, only three methods are available to so regulate births: (1) the prevention of fertilization through abstinence or interference; (2) the termination of pregnancy; and (3) by infanticide. Leaving aside the first one as highly unreliable, that leaves only the last two as sufficiently effective methods. Whatever the largely irrelevant attitude of men to these last two options may have been at any given time and place, *all else being equal* these are not options a woman is likely to willingly choose in the absence of a compelling reason. The fact that a woman would choose to employ a method of termination or accede to the employment of infanticide suggests that in most cases environmental pressures—of which competition is a component—were prompting such actions. So, once again, the facts offered as being contradictory of "population pressure" are actually a measure of it.

Lastly, often lost amidst the criticism of the "prime mover" aspect of Cohen's (1977) thesis is the ancillary point he makes to the effect that the habitable world—including previously uninhabited parts of it—had been largely occupied by our species in the estimated ~60,000 years between our departure from Africa and the end of the Pleistocene. Population had clearly grown substantially and there is no plausible reason to think that such growth would slow down in the absence of Malthusian penalties promoting a slowdown.

As argued elsewhere (Rosenberg, 1998) and ably demonstrated by Richerson et al. (2001), population pressure is a very real factor in the evolution of culture and was a major factor in the evolution of food-producing societies. The real problem with these early population pressure models was, as also noted by critics, with their mechanics, which held it to be a "prime mover" that in and of itself produced the outcome being explained. Competition, whether one chooses to call it "population pressure" or not, is not (contra the early population pressure models) a "prime mover." Competition is simply an environmental force that does nothing but trigger purposive behavioral responses to its highly variable manifestations in accordance with the variable multiple affordances available for engagement with, while simultaneously also being one of the forces selecting from among the affordances being engaged with and the manner in which they are being engaged with.

Competition and Biased Responses

Unfortunately, the term population pressure carries the very negative baggage of its initial simplistic usage. Moreover, as thus far used for explanatory purposes, it is only one manifestation of competition (competition for food resources). For those reasons, going forward, the more broadly meaningful term competition will here be used in lieu of it. However, to deny a pivotal proximate role in generating cultural novelties to what was called population pressure (i.e., competition), in addition to its ultimate role as a driver of natural selection, is tantamount to denying competition's pivotal role in evolution, including cultural evolution. And, as noted above, to deny the ubiquity of competition is to deny Malthus' insights and to deny Malthus's insights is to deny the mechanics of Darwinian evolution. People will respond to competition in purposeful ways, some biased by human behavioral tendencies. It is these tendencies that bias proximate causation and in the process they can potentially trigger punctuational episodes by causing the sorts from the organic hierarchies to overpower the sort from the information hierarchy within the adaptive hierarchy.

In one of its expressions, competition simply drives selection. In the other, in sentient species it will tend to bias the appearance of behavioral novelties in favor of ones that confer some potential advantage in that competition. But, competition cannot be said to proximally cause such advantageous novelties to appear; competition merely triggers the initiation of purposive behaviors intended to address that competition. The biases do that by mediating the competitions in the organic hierarchies and behavioral selection in the adaptive hierarchy to somewhat favor the appearance of advantageous novelties.

To think of competition, whether as what was called population pressure or as any of its other possible context-dependent manifestations, as a "prime mover" in cultural evolution is the equivalent of considering selective forces to be "prime movers." Thus, early population pressure models were at once incomplete for omitting proximate causation and, more importantly, also anthropologically meaningless for being wholly self-referential. They were simply descriptive statements of the presumed-to-be obvious, corrected only partially by later properly evolutionary models by including references to the selective processes that produced the outcome being explained. They all omitted how and why the behaviors in question came to be practiced in the first place for selective forces to act on because cultural evolution cannot count on the analog of random mutation to provide the answer to either of those two questions. Meaningful cultural innovations are not randomly produced primarily by the behavioral analog of some N-number of proverbial chimpanzees pounding on typewriters and only very rarely producing a work of real prose's analog in the form of an improved behavior or technology, which can then be selected for. Nor can rational choice be counted on, because that ignores both the cultural filter and cultural constraints, not to mention behavioral biases.

Just to be clear, competition's role in proximate causation should not be viewed in absolute terms; nurture (i.e., the information hierarchy) does play a role in behavioral selection and more often than not a very large role. The proper way to view competition is to consider it the default trigger, the default hypothesis so to speak, that must be rejected on the basis of data in order for an alternative explanation to be an acceptable replacement. In other words, as long as the context-dependent competitive tendency toward a particular behavior is well documented *by any scientific discipline* and some evidence for it is available, the order of proof needs to be reversed given competition's central role in evolution. Rather than requiring some measure of proof of competition and its consequences for it to even be considered for a role in the process being investigated—the basis for the hypothesis's relative privilege being the well-documented tendency itself—the proposed alternative should require, at a minimum consilience with *everything* else we already know regarding behaviors, etc. that relate to the phenomenon in question, for it to be considered as a possible replacement for competition in the explanation for what *triggered* the behavior in question.

Lastly, it is important to keep in mind that people both individual and collectively can and often enough do willingly act in ways that harm their interests as individuals in one hierarchy in order to further their interests as individuals in one or more of the remaining hierarchies, especially since doing do so may very well be seen by them—correctly or incorrectly as the case may be—as a roundabout way of furthering those same harmed interests or furthering other interests that they calculate to be a priority. Metaorganically promulgated, but organically maladaptive behaviors can persist for a relatively long time before selection in the organic hierarchies weeds them out: the aforementioned Shakers were a dynamic group from their founding in 1747 until well into the late mid-Nineteenth Century—almost the same duration as the Native American Great Plains way of life—and in decline for over another hundred years after that. That span is certainly long enough for them to be an archaeological entity visible to future archaeologists, who would be wise (if they did not already know the opposite to be the case) not to fall into the adaptationist trap and attempt to explain their way of life as a product of selection in the organic hierarchies. The point is that neither the existence nor structure of that specific set of nested cultural and associated social entities can be explained by reference to adaptations favored by selection in the organic hierarchies. Even relatively long-lasting cultural and social entities can simply be one-off drift-produced flukes-of-culture; failed experiments produced likely as not by chance novelties produced within the information hierarchy as opposed to ones triggered by the sorts from the organic hierarchies driving behavioral choices. The key to distinguishing the two sources is the presence or absence of pattern across a group level, which among other things (e.g., similar affordances) is arguably a product of behavioral biases.

Conflict Avoidance, and Other Biased Responses to Competition

Competition, whether over resources or any other desirable thing, did not begin with the end of the Pleistocene and the dawn of food production. On a long-term basis it, was been constant and ongoing for as long as our species (or any other species, for that matter) has existed, though it has been variable in its manifestation and expression and it continues into the present. Europe's estimated population more than doubled during the Medieval Climatic Optimum before crashing due to famine and disease/plague (i.e., Malthusian penalties) with the onset of the little ice age at the beginning of the fourteenth century. More importantly, for present purposes, competition has produced behavioral adaptations, reinforced or undermined by case-by-case conscious calculations, for dealing with competition, such as a tendency for groups to fission as a mechanism for conflict avoidance, territoriality as a mechanism to maintain access to resources in the face of competition when fissioning is hampered, and display for discouraging competitors from more intensely competing, etc.

No one of consequence seriously believes that some Upper Paleolithic band of foragers in central Asia one day just decided to uproot themselves and travel several thousand miles to eastern Siberia and its continuation as Beringia on the offhand chance there was an undiscovered continent waiting to be discovered at the end of that journey, or because they were following a mammoth herd that had itself inexplicably decided to migrate several thousand miles to the east, or because the harsher environmental conditions to the northeast were somehow more alluring in and of themselves. Rather, socioeconomic processes, in particular competition within groups[8] (whether it was over resources, mates, real estate within the camp on which to build a shelter, or any other even more petty dispute) and the resultant cumulative resentments and disagreements, amplified over time and by the overall level of both intra- and intergroup competition, tended to cause groups to fission. More often than not this was because, to one faction in the dispute, the increasing costs of the dispute outweighed the cost of just leaving for an unoccupied area (probably already explored to some degree) just far enough away[9] as to avoid having to tangle with the other faction ever again. In other words, areas to the east were likely more alluring only in the context of then-current social circumstances and their broader environmental milieu—for example, more game to the east due to less

[8] Competition between groups starts with intragroup competition, because the resultant fissioning is what produces competing groups.

[9] Habitable areas to the west, north, or south would likely have been already occupied and moving in those directions to join other groups or carve out a space for themselves in their midst would have just substituted another competition for the one being terminated by means of avoidance and thus usually pointless over the long run. This is because it would have just postponed the dynamic for some relatively short length of time and then reinstituted it within the group so joined or with the groups elbowed somewhat aside to make space for themselves.

competition for game to the east. Multiply that process by some large number of times involving intermittently different groups and you end up after some (surprisingly short) time (see Richerson et al., 2001 for the math) with groups living in Beringia, eventually in Alaska, and eventually in Tierra del Fuego.

The same dynamic likely applies to the move(s) by our species out of Africa and the still earlier tentative moves to exploit more open country by the distant ancestors of our species in the face of intensifying competition perhaps due to habitat loss. To believe otherwise is naïvely teleological, the absence of suitably concrete (in the eyes of critics) evidence of "population pressure" in the form of violence notwithstanding.

When seen as feasible, such competition-based, conflict-avoidance fissioning is usually the least risky strategy in the face of rising competition-induced tensions and is well documented ethnographically (Lee & DeVore, 1968: 9; Lee, 1972) and primatologically (e.g., the Kasakela-Kahama conflict of the mid 1970s at Gombe). In that primatological instance, the Kahama group simply could not get far enough away from the Kasakela group while also staying within the relatively safe (from humans)[10] confines of the park to allow the competition and resultant violence to wither away instead of intensifying. Moreover, the use of fissioning by both humans and chimps as an attempt to forestall more intense and perhaps violent intragroup competition by both chimps and humans strongly suggests that it is a default conflict-avoidance response in humans to intensifying competition, as long as it is seen as feasible and the *apparent* cost-benefit calculations do not make fissioning appear to be a losing proposition in either absolute terms or relative to other purposive options. If they do, other affordances that appear to offer at least the possibility of relatively better outcomes become increasingly attractive.

Other factors may or may not have been involved in specific instances of fissioning, but local intra- and intercommunity competitions in one form or another were very likely more or less a constant. At best any other potential factors were secondary considerations affecting specifically where to go and/or how to get there, affecting the specific form the fissioning or whatever form the chosen response took, but the proximate cause was the social manifestation of intragroup competition and the biased response was fissioning if possible.

More importantly, competition over resources can trigger other biased responses, particularly if the colonization of new areas through fissioning is not an easy option in that competition. Of these, the ubiquity of territoriality in the animal kingdom clearly indicates that it is an effective method for dealing with competition because that very ubiquity in nonmigratory species indicates it was selected for in a myriad of different contexts and thus an innate response to competition in a socially circumscribed environment that maximizes reliability of access to anything that territory is a reasonably dependable proxy for. Territoriality is also often associated

---

[10]The history of that conflict indicates that in terms of objective reality they calculated poorly and probably would have been better off taking their chances with human poachers in areas outside the park.

with the tendency to display. Display serves to demonstrate both occupancy of the territory, the willingness of the occupants to defend that occupancy, and often incidentally the strength of that potential defense.[11]

Such innate behavioral biases (e.g., favoring conflict-avoidance fissioning, territoriality, display) were fixed in the organic hierarchies. They feed from there (the organic hierarchies) to the adaptive hierarchy where they bias behavioral selection in the context of the framework for and constraints on behavioral selection imposed by the cultural systems of the information hierarchy (CSs $\rightarrow$ BS). But, these are just biases. It is worth repeating that people are individuals and individuals are not prisoners of their genes. Nurture, in the form of the information system one operates under, always also plays a prominent role in behavioral selection. To say that a behavior is biologically biased to occur is not the same as saying it is biologically determined at either the individual or group level.

Proximate Causation and Childe's Revolutions

Two relatively recent prehistoric developments, one technological and the other social, stand out both for their importance in the evolution of now widespread cultural institutions. Both are proximally the product of constrained contingent processes rooted in biased behavioral selection and both exhibit a high degree of parallelism/convergence because of those behavioral biased purposeful actions. Childe (1936) referred to them as the agricultural and urban revolutions. The first of these refers to the evolution of the first sedentary food-producing societies at approximately the end of the Pleistocene and the other to the subsequent contingent evolution of very large, typically socially complex, formally hierarchical, and increasingly nucleated human societies made possible by the surpluses that food production allowed to be produced.

The evolution of both, in all their varied trajectories and specific forms, was arguably a product of biased behavioral choices triggered by competition and serve as good examples of how the processes that constitute proximate causation operate as regards cultural evolution by driving behavioral selection. Moreover, the evolution of both required purposeful innovative behaviors that are anathema to the kinds of cultural systems that constitute the starting states from which they evolved: a departure from generalized reciprocity for the beginnings of food production and, at the most basic level, a departure from (even just nominal) egalitarianism for the evolution of the most ultimately successful[12] forms of sociopolitical complexity. Thus, both also serve as examples of states produced by some variable number of punctuational episodes.

[11] E.g., the number of voices in a Howler monkey troop's main vocalizations.

[12] Measured in terms of societal size.

Chapter 8
Proximate Causation and Pattern Part I: The Paths to Food Production

"History doesn't repeat itself. But it does rhyme."
Mark Twain

But What About Behavioral Selection?

Until the last third of the twentieth century, the evolution of food production and settled village societies were largely viewed with implicit reference to the *telos* of progress and the assumed economic superiority at each step of the new over what it replaced. In that view, the development of food production and associated settled village societies simply awaited discovery and/or suitably conducive cultural conditions. The requisite behaviors for the transition were simply assumed to begin being practiced once discovered because of their recognizable superiority. With the shift to causally oriented models in the last third of that century, in the wake of studies that cast doubt on both the need for discovery due to a lack of prior knowledge and the absolute superiority of food production over foraging (e.g., Boserup, 1965; Lee, 1965, 1968), two broad types of causal explanations were offered.

The first type of explanation was essentially adaptive. In these, human functional awareness and adaptiveness largely replaced progress as ultimate causation—food production evolved to feed people better in the context of need because providing food is what food production primarily functions to do. Cultural evolution was thus still viewed largely as a self-contained process of transformation, but now rooted in the human capacity for rational thought. The only difference from the preceding progressive models was that cultural 'evolution' was now essentially seen as a process of adaptive problem-solving in which the "problem" and its "solution" were assumed to be obvious. In the case of earlier progressive explanations, the transformation produced an obviously superior state of cultural being for the groups so transformed—a teleological step toward the distant and far superior present. In the case of these adaptive explanations, human groups intentionally transformed themselves for the specific purpose of optimally adapting to a given set of environmental conditions. Natural selection, if invoked at all, merely served as the a posteriori

M. Rosenberg, *The Dynamics of Cultural Evolution*, Studies in Human Ecology and Adaptation 12, https://doi.org/10.1007/978-3-031-04863-0_8

justification for the continued existence of the transformed state—to rubber-stamp the transformed state's function. The transformed state's function was viewed as inherently adaptive relative to the proposed environmental conditions that triggered the transformation in the first place, be they population pressure or climatically precipitated sudden resource availability.

In most cases, for these adaptive models, cause and optimal response were implicitly assumed across the board to be transparently relatable; and, thus cause inevitably produced the effect being explained. The proximate and ultimate causes of the transformation were thus one and the same. In essence, such explanations were (environmentally) deterministic. Chance entered the equation only contextually and only to the degree that any given shift to food production was said to hinge on circumstantial contingencies, such as the occurrence of environmental triggers (e.g., availability). Find the critical environmentally produced affordance and one finds the 'cause' because the opportunity and trigger are one and the same. Implicit is the assumption that people would have invariably engaged with the particular affordance that was food production or sedentism (depending on which was the cart and which the horse in a given model) so long as it was available and as soon as it was available to engage with. The environmental trigger was thus not just a necessary condition, but in and of itself also a sufficient condition to produce the new (adaptive) cultural system.

Though adaptive models were abstractly agential in orientation, cost-benefit calculations and behavioral choices (i.e., behavioral selection) based on them were generally still explicitly not taken into consideration, even though they were precisely what cast doubt on the preceding progressive models. Disregarded virtually across the board by these adaptive models were Boserup's (1965) conclusions that intensification is labor-intensive and Lee's (1965, 1968) conclusions demonstrating that foraging yields sufficient food even in relatively harsh environments such as the Kalahari. Add to that the growing awareness that increased sedentism carries significant risks such as resource depletion (e.g., Kelly, 1983) and there are ample reasons to conclude that people would not avail themselves of the proposed affordances, even if they were an option, in the absence of additional considerations encouraging them to do so. Thus, while the existence of the proposed affordances was theoretically a necessary condition to produce the outcome being explained, in and of themselves the existence of the proposed affordances was not a sufficient condition. Missing was any reference to proximate causes triggering the requisite behaviors that would tip the cost-benefit equation in favor of a people engaging with the affordance(s).

The second type of these causal models invoked social processes, most recently revolving around religious edifice construction. These were offered as general alternatives to adaptive models and typically revolved around social competition either explicitly or implicitly. However, while also abstractly agential, these failed to accommodate cultural constraints on behavior, such that the requisite behaviors (e.g., goal-oriented social manipulation by would-be alphas) proposed to have led to food production would likely have had the opposite of the desired effect in the initial stages of the process or come into functional play only once food production was already established. That is, they revolve around overt social competition and associated social manipulation, behaviors that are anathema in documented examples of the simplest mobile foragers (e.g., Boehm, 1999). Such behaviors are thus subject to intensely negative cultural selection in the very cultural systems that are generally considered the approximate socioeconomic starting conditions for the process.

Moreover, if social sanctions do not successfully quash such behaviors, people will simply opt to leave due to the demands inherent in such manipulative behaviors becoming burdensome, unless there are demographic constraints that prevent people from doing so. In other words, locational constraints are a precondition for the operation of such models. Simply put, at best such social models require locational constraints (i.e., competition and territoriality) to become operative, the very proximate causes to be proposed here, and at worst they require the very conditions typically set in place by the beginnings of food production and settled village life (e.g., surplus production and material inequality) to explain the beginnings of food production and settled village life. Social processes of the type proposed, in and of themselves, only potentially work in conditions where surplus production is already a given, as in societies that are already producing food or in surplus-producing complex hunter-gatherers. In that latter case it apparently led to settled village life and often a significant measure of social complexity in the absence of significant food production until very much later, a trajectory I will return to later.

A more recent third "generation" of models for the origins of food production have brought evolutionary forces to the forefront of the processes involved in the origins of food production and settled village life (e.g., Rindos, 1985; Richerson et al., 2001; Kuijt & Prentiss, 2009; Zeder, 2012, 2016). However, all only satisfactorily address the issue in terms of ultimate causation, still glossing over proximate causation (i.e., behavioral selection), and implicitly taking it as a given that people will engage with the given affordance(s) presented to them despite the human capacity for calculating costs and benefits and what those calculations imply about the likelihood of engagement with any affordance that is more labor-intensive than the present case. The problem once again is that, all else being equal, the math concerning intensification at any single point in time simply does not add up to a net benefit at all in the absence of other considerations that would modify the math. Models revolving around niche construction, to take just one example, require behaviors that successively modify a niche. From whatever starting point one chooses to begin examining the process, at some number of points in the progression to food production those niche-modifying behaviors are riskier and/or more labor-intensive than what they replace, but they are presented as assumed to have been engaged in despite the likelihood that they would not be in the absence of other factors promoting their practice. Invoking climate change as that factor (e.g., Richerson et al., 2001; Kuijt & Prentiss, 2009) is all well and good, but that requires one to further state why that should be so and they generally do not take that next step except to sometimes fall back on availability. Richerson et al.'s (2001) invocation of population pressure is the exception, but it goes no further than proposing function as (ultimate) cause in the same manner as did the earlier adaptive models.

All such evolutionary models (and a fair number of the earlier adaptive models) are internally coherent and typically make reference to accepted understandings concerning either the physical environments that served as the context for the models' mechanics or the mechanics themselves. But, like the earlier adaptive models, evolutionary models make no explicit attempt to fit themselves within the broader domain of what we know of human behavioral tendencies, the social context for culturally patterned behaviors (other than perhaps with reference to cultural selection and transmission), and economic behaviors in particular. That is, they beg the question of how can you get selection for a behavior under the normal cultural and ecological conditions that constitute critical aspects of the mobile foraging starting conditions, if there are plausible reasons for the behavior to never be widely engaged in due to its labor intensiveness, and thereby become visible to the operation of either cultural or biological selective forces. In other words, population pressure invocations aside, they beg the question of why people would choose to begin working harder, whatever the long-term selective benefits in the organic hierarchies might be—for organic modes of selection only operate in the present—when the lower level of work they were doing up until that point was sufficing.

Adaptive and more current evolutionary models for the origins of food production, whether coevolutionary (including their current incarnation as niche construction) or attraction models (including their current incarnation revolving around sustainability), do not address the likely negative behavioral selection generated by the relatively labor-intensive nature of even the simple exploitation (including processing) of more *r*-selected as compared to more *K*- selected resources, not to mention the labor intensiveness of actual food production. Moreover, sustainability implies concerns about the possible depletion of the focus of such activities in the absence of efforts to so sustain it, for if possible depletion is not a concern why bother with efforts to sustain it, which leaves the further question of why depletion is still a risk if movement could address that risk at an arguably lower cost. Likewise, as explicitly noted by Kuijt and Prentiss et al. (2009), niche construction is not an enclosed ecological tunnel that, once entered, has no behavioral exit before its ultimate end. The thing is that there is never at any point along the way only one affordance available to engage with and the decisions concerning which to engage with are made purposively. That leaves the questions of: what is the purpose of engaging with a new one that is more labor-intensive than the one currently engaged with, and why are less or equally labor-intensive affordances not available to engage with.

Yes, *r*-selected species are said to yield more, are generally easier to manipulate, and/or are more difficult to deplete, but these models do not satisfactorily address the question of why that should matter "now" when it did not matter "before." They are all implicitly predicated on the assumption that people do not attempt to calculate some measure of current cost-benefit as regards intentional actions because the very existence of that math is never brought into the discussion. They all proceed as if natural selection is the only force operating to produce economic behaviors, negative cultural and/or behavioral selection never exists, and cultural transmission inevitably follows natural selection in passive fashion. That is, they proceed as if the selective flow is unidirectional in the form of NSs → [BS/BSs] → CS and direct from NS to CS. This is largely because they are conceptualizing culture in the wider sense as encompassing both information and behavior and thus always of implicit necessity operating in tandem. By failing to distinguish between BS and CS, they leave no option or mechanism for them to work at cross purposes. They thus fail to account for any possible negative behavioral selection (produced by CSs → BS, not to mention by CSs → NS),[1] due to either the labor intensiveness of the innovation and/or the constraints of Bauplan, potentially overpowering any natural selection favoring and thus fostering the behavior. In other words, they all lack a reference to the details of proximate causation, the context and motivations for the required innovative behaviors, and take the requisite labor-intensive behaviors as a passive given, with no attempt to address why that should be so beyond perhaps general references to climate change and in the case of Richerson et al. (2001) also population pressure.

To repeat: if a behavior is not being engaged in, it cannot be transmitted culturally because there is nothing to transmit by observation and nothing for selective forces in the organic hierarchies to operate on. Why engage in an intensificatory economic behavior unless there is some perceptible current benefit that is sufficient to overpower the latent negative behavioral selection generated by the labor intensiveness of that behavior, such that the *current* benefits of that economic behavior outweigh the added labor costs? The fact that a behavior potentially has selective value does not mean the behavior will come into existence to be selected for in the context of present conditions. It merely means it will be selected for if and when it does. The point is that environmental models, evolutionary or otherwise, are not necessarily wrong in any absolute sense; they are simply incomplete. The same can be said of social explanations. In that sense, these various kinds of models are all at best elegant semicircular pegs in a field of round holes, fitting in but not fully filling the holes they are proposed to fully plug and requiring the complementary semicircle, so to speak, to be filled in where possible by the incorporation into the model of an appropriately fleshed-out (i.e., locally relevant) version of proximate causation. For the beginnings of food production and settled village life that complementary semicircular peg is locational constraint and its variable consequences.

[1]The ability to calculate is fixed by differential reproduction in the reproductive hierarchy, based on what the sort from the economic hierarchy provides (i.e., good "mathematicians" do better economically). The formulas for calculating are a feature of the information hierarchy and the sorts flowing from that hierarchy to the others.

Territoriality and Constraint

So long as a set territory is a reliable proxy for critical (often economic) resources,[2] animals have an innate tendency to be territorial in the context of the Darwinian given that is competition for those resources (Maynard Smith, 1982; Barash, 1977). Territoriality reduces the day-to-day need to defend resources from being poached by other entities through signaling that tells others that the territory is being actively defended and that the attempted poaching of resources risks a potentially violent confrontation. In the broader animal kingdom, this defensive threat is often in the form of calls, but it can also be in the form of physical displays, all designed to notify would-be poachers not just that the territory will be defended, but that it will be defended by one or more intimidating defenders.

When an active defense turns out to actually be required, territories also provide an advantage to the current holder as the defender of that territory (e.g., see Etkin, 1967; Van Sommers, 1972; Barash, 1977). Specifically, the cost-benefit equation regarding potential escalation of the competition favors the defender, who is likely to be more willing to endure the costs of escalation because they have more to lose than does the intruder (e.g., see Maynard Smith, 1982). The costs of the competition increase as it escalates and include the risk if not likelihood of injury or worse if it escalates to violence. The objective cost of escalation is the same for both the owner and the intruder, but the potential loss of benefits means the owner has more to lose than the intruder does by abandoning the competition, making the owner more willing to bear the cost of escalation. That is, if the intruder breaks off the competition in the face of a vigorous defense (and thus uncertain outcome) before the costs become significant, much less existentially threatening, then nothing is gained for the initial attempt's relatively minor cost investment, but nothing is lost besides that minor investment either. The intruder is not significantly worse off than they were before the gambit and the option is always there for them to go somewhere else that is unoccupied and thus undefended, albeit probably poorer, or try again elsewhere and hope for a weaker defender next time. However, if the owner breaks off the competition at the same stage, they lose their territory, making them materially worse off, which typically makes them more willing to bear the higher cost of escalated competition than is the intruder. Add to that the possible defensive advantage to the owner of familiarity with their surroundings and the costs potentially shift even further in favor of the territory's current holder. In humans, this defensive advantage is further amplified by the use of ranged weapons, which allow even an undetected lone defender to ambush a larger group, inflict injury and still escape undetected to raise an alarm (e.g., see Kelly, 2005).

Territoriality requires competition, for without competition for resources there is no reason, beyond perhaps some degree of logistic convenience, for any group to

---

[2]Or those resources are in turn a reliable secondary proxy for some other critical (typically reproductive) resource for which competition also exists, such as potential mates attracted to those resources, as in stretches of beach for elephant seals.

restrict its movements to a given area insofar as that and neighboring areas are nonexclusive. But, in that case the area in question would constitute the entity's home range and not its territory. Territories are in principle at least nominally exclusive to a group (though group membership does not of necessity have to be) and when necessary defended. In contrast, ranges overlap and are neither exclusive nor defended, even though individuals may nevertheless be mutually hostile when they encounter each other in areas where their ranges overlap.

Territoriality is not some absolute and unvarying sociodemographic construct. All else being equal, the greater the competition for resources, the more likely that ranges evolve into territories, the more rigidly defined territories are likely to become, and the more vigorously a territory is likely to be defended because the possibility of successfully establishing a comparable replacement for a lost territory is commensurately reduced. Among human groups, territoriality varies considerably more in its manifestations that it does for other species because of the complexity of human sociality and the fact that human groups are usually not social isolates. In other words, human territoriality is as much a social as an economic construct (e.g., see Bettinger, 1999, 2001). Within any given social environment, a given individual will normally have friends/allies and others who are some degree of neutral to hostile. The same applies to groups as individuals. Thus, territoriality in humans, unlike most other species, is not a case of one against all. It only requires potential defense against individuals who are not friends/allies, because one key basis for the friendship/alliance is mutual respect of respective rights, including territorial rights.

At the most basic level, territoriality can range from loosely fluid to rigid, varying with the degree of competition (see Dumond, 1972; Charles & Buikstra, 1983). However, even so it will always reflect socioeconomic and sociopolitical realities and can range from being completely exclusive to nonexclusive with respect to select other groups considered, friends, allies, kin, or just fellow tribal or societal subgroups. Such reciprocal social arrangements with other groups obviate the day-to-day need for active defense with respect to those groups, in that they both maintain rights to territory by mutual agreement while also simultaneously still allowing for a degree of cooperation, including perhaps cooperative defense, as well as the movement by individuals between groups as necessary for successful reproduction, dealing with short-term local resource scarcity, etc.

Territoriality requires that a territory be a reliable proxy for something worth the cost of defense. For economic resources, it requires that the resources contained within it be both predictable and sufficiently numerous to reliably provide for the territory's holders (see Dyson-Hudson & Smith, 1978; Layton et al., 1991). The greater that predictable availability, the more practical territoriality becomes, the more valuable a territory becomes, and the more likely it is to be defended if necessary because it is increasingly worth the cost of defense.

Within a given territory, not all the exploitable locales yield equally desirable or efficiently exploitable resources, not all locales yielding even the same resources yield them in equal measure, and even the same locale will not yield the same resource in the same measure all the time. For mobile foragers, that variability is addressed through mobility. However, territoriality has the negative effect of constraining that mobility and thus limiting access in case of need to alternative locales outside the territory and the resources that those alternative locales could potentially yield. Given the tendency of populations to grow over time until checked, restricting a group to only a given territory's resources creates the rising risk of resource depletion over time within that territory as Malthusian checks are approached, and the smaller is that territory the greater is the risk. Leaving innovative resource extraction behaviors momentarily aside, where possible, this risk is mitigated by mobility and the alternative locales within the territory yielding either the same or lower ranked alternative resources which mobility makes available for exploitation as needed. Such mobility may be by a group or its logistical subgroupings within its territory. However, the smaller the territory, the more such mobility is constrained. Alternately, it may be mitigated by individuals and fissioning-produced splinter groups moving between territories in the case of reciprocal social arrangements[3] that allow for the movement of people between territorial groups. Unfortunately, the greater is the competition, the greater is the risk that such social arrangements will break down due to group-level conflicts of interest.

Competition and Territorial Compression

The degree of constrained mobility produced by territoriality is fundamentally a product of territorial size because territorial size bears on the degree of mobility potentially available to a group within its territory. All else being equal, a large territory allows for a relatively high degree of mobility within the confines of the territory by simple virtue of the available space, while a smaller territory allows for commensurately less mobility. All else again being equal, a larger territory will also likely encompass a wider set of potential resources and a larger number of resource-yielding locales than a smaller one. Thus, all else being equal, the larger the territory, the more mobile the group occupying it can be and will tend to be. This is because the territory allows it and because it is the more risk-free behavioral pattern.

All else once again being equal, the larger a given individual territory is, the more the risk of resource depletion is likely to be mitigated by the encompassment of either more locales for any given resource or of other locales yielding alternative resources; the smaller the territory the more that risk is increased. On the other hand, all else being equal, the larger the territory the more difficult and energetically costly it is to defend effectively *in its entirety* should the actual need to do so arise. This is because one cannot be everywhere at once within a given territory and the more locales that constitute the most important elements of "everywhere" the more of those locales become relatively poorly defended. This applies particularly to the more distantly peripheral locales within the territory that yield lower ranked

[3] A competitive environment does not forestall cooperation between some groups while they also more actively compete with other groups.

alternative resources not regularly exploited, which therefore tend to not be visited as often as other locales, and whose defense thus benefits less from the mere fact of the group's physical presence or even proximity. Thus, squatters can more readily move in to exploit such infrequently visited peripheral locales without risking a direct confrontation and the potential costs such a direct confrontation would entail.

All else yet again being equal, the more intense the competition for resources, the more frequently territorial defense will actually be required. Thus, given continued competition, for each group and the specific makeup of its territory there is some theoretical uniquely local threshold level of competition at which the perceived cost[4] of defending productive-but-distant and/or marginally productive portions of a territory begins to outweigh the perceived costs and risks inherent in the exploitation of a more spatially restricted territory.[5] If and when that happens, cost-benefit calculations will eventually tend to favor a retreat from the more distant, less important, and/or more difficult to defend areas. This will produce the compression of what had previously been larger territories and the concomitant creation of new territories encompassing the made-available abandoned spaces to produce a larger number of groups in a larger number of smaller territories (see Rafferty, 1985: 120). The degree of compression and associated packing that can actually occur in a given situation is obviously limited by both the available resources within existing territories and the extractive technologies available to the groups in question.

Given a system of territories, it is a further given that not all territories will be equally productive and thus capable of supporting similarly sized groups. Given continuing if not escalating competition, it is theoretically inevitable that a certain percentage of the population will develop into non-territorial splinter groups. Such splinter groups are theoretically a product of group fissioning, the propensity for which logically rises with the stress levels within a group, as would result from resource scarcity relative to group size (e.g., see Lee, 1972). To the degree that such stresses are not confined to just the group that first fissions, the process will tend to eventually repeat within whatever now-bigger group(s) the departing segments move to join. That is, to the degree that they are then relative newcomers to any group they end up joining, breakaway groups are likely to be less well integrated within the group they join than longer standing members. Thus, if and when that group also fissions due to its now-larger size, it is logically more likely that the relative newcomers will again tend to be the core of any new splinter group instead of other, longer-standing and better integrated members of the group that they joined. If this process repeats often enough, such splinter groups can potentially become permanent "floaters."

Such floaters might sometimes join with established groups if conditions permit. At other times, they might exist independently as splinter groups—the distinction between the two ends of the continuum that spans splinter and logistically motivated subgroups of larger groups is the regularity with which they consistently rejoin the

---

[4]This may or may not track actual cost.

[5]This may be dependent on the technology to actually do so.

same larger groups. In that social state, floaters could subsist with or without permission on the resources available in those parts of existing territories that are not being exploited in the current year by the holders of that territory. In essence, these hypothetical floaters are called for by normal hunter-gatherer group aggregation-dispersion dynamics and Malthusian demographics taken to their logical conclusion. True, the actual existence of such floaters is poorly documented, but then no one has given much thought to look for them. Moreover, their existence is hinted at in some of the hunter-gatherer literature. For example, to the degree that the documented Basin Shoshonean groups each had their own territories, other than neighboring groups looking to expand their territories, which is not mentioned as having been the case, who else but such floaters would e.g., Owens Valley groups tend to have been driving off their pinyon stands with displays of nonlethal aggression (see Steward, 1938), or (nonindigenous intruders aside) Andaman islanders off their territories with lethal aggression (see Kelly, 2005)?

Such floaters represent potential competition for holders of existing territories (see Barash, 1977: 271). To the degree that existing territories, particularly larger or more productive ones, are compressible in the context of an existing or otherwise known extractive technology which potentially allows the extraction of sufficient resources from a smaller territory, such floaters may succeed in carving out new territories. Such new territories would likely be at territorial interstices from the most distant and thus difficult or costly to defend portions of territories held by other groups, who find the cost of increased subsistence-related labor to be outweighed by the costs of defending such marginally important locales. While there are limits, dependent on extractive technology, to such territorial compressibility, the threat of such attempts at territorial insertion is a constant, as is the potential threat of further encroachment by either groups of floaters or one-time floaters who are now de facto holders of only marginally productive territories. The most immediate day-to-day competitors for territorial holders are thus not so much long-time established territorial groups as such nascent groups of floaters who hold no firm territorial rights to speak of, or associations of one-time floaters who have already managed to carve out marginally productive territories at abandoned territorial interstices.

The point is that, for a mobile forager group experiencing competition, the danger is not so much of being displaced from its territory as a whole, as it is of being displaced from one or more locales within its territory and the greatest danger is that of being displaced from one or more of the most productive locales. The point to bear in mind is that a group of floaters squatting at a locale are in a very real sense in possession of that locale at that point in time, negating some if not all of the defensive advantages otherwise held by the group on whose territory they are squatting; and, the longer they have been unchallenged in so squatting, the less of a disadvantage they are at in a confrontation with the group on whose territorial perimeter they are squatting. The basic problem confronting a mobile hunting-gathering group experiencing such competition for locales within its territory stems from the fact that it can only physically occupy—and thus most effectively and efficiently defend, or by their mere presence/proximity likely forestall the actual need for physical defense—one locale within their territory at a time. The potential problem, should it come to that point if the competitive environment requires a choice to be made, is one of deciding which locale(s) within an existing territory to relinquish because they do not pay to defend and which to commit to continuing to defend.

The various productive locales within a given territory are not all equally valuable for subsistence purposes, even in the case of those that yield equally desirable resources. The first and most basic aspect of that value is the locale's predictable productive capacity as regards its most important resources in the context of a given extractive technology and a given limit to the daily travel time willingly accepted for resource-gathering forays directed at those resources. The second would be the locale's greater productive capacity, which also includes the potential productive capacity of all other locales that could, if necessary, also be effectively (if less efficiently or desirably) exploited and defended from the primary locale without leaving the primary locale potentially subject to loss. Degree of resource diversity within the locale's vicinity would likely also be an important consideration, with high diversity within the catchment area of the locale and its potential major satellites adding value by lowering year to year resource risk in the form of containing lower-ranked exploitable alternatives available for times of need.

This relative locale value is the ultimate basis for any potential decision regarding which locales to abandon and which to defend. It is in this context that the cost-benefit calculations cause specific resources to exert an attraction or "pull" of the type proposed by the various evolutionary and adaptive models on a group contemplating decreased mobility. That is, competition for resources makes certain resources more desirable than perhaps they had been previously because through added labor they can compensate or be made to compensate for the reduced mobility inherent in territorial compression.

Territorial Compression and Innovation

What ultimately determines whether a reduction of mobility will actually occur in the face of competition is the availability of an option to engage in innovative subsistence behaviors that would make survival within a spatially reduced territory both reliably possible given the nature of the resources contained within it and less costly than continuing to defend the territorial status quo. In the absence of such an innovative behavioral option, reduced mobility due to competition will tend not to occur because the benefit of continued defense outweighs the cost of such defense. The necessary innovative subsistence behaviors may simply be a repertoire of known behaviors or technologies hitherto not normally utilized in the manner now contemplated, or they may be entirely new behaviors or technologies based on preexisting or newly discovered knowledge. Whatever their source, they have to be perceived to be applicable in the group's environmental and social situation or the option to so innovate and engage with the affordance will not be considered practical. Whether or not it actually is practical in some objective sense is another matter entirely, but the relative relatability of cause and effect in the case of subsistence behaviors suggests that more often than not attempted innovative subsistence behaviors will have some degree of practical (and hence selective) value.

As long as there are no cultural constraints operating in opposition to the contemplated innovation(s), the costs of abandoning the status quo will be lower than if such constraints do exist, as the social disapproval for violating cultural norms is very much a cost factor; and, the more those constraints are integral to the cultural Bauplan, the higher the social costs of violations. Thus, the more a contemplated innovation is in violation of the cultural Bauplan, the more motivating the competition-produced sorts from the organic hierarchies need to be in order to overpower the sort from the information hierarchy and trigger the behavioral innovation despite social disapproval. Given the existential nature of the selective forces operating in the organic hierarchies, the sorts from those hierarchies certainly have the potential to overpower the sort from the information hierarchy as regards behavioral selection in the adaptive hierarchy if the competition is sufficiently intense.

All else being equal, making do with a smaller part of a hitherto larger territory requires the more intensive exploitation of the now smaller territory's resources than would have been the norm were the previously larger territory still fully available to exploit (e.g., see Wills, 1988: 34ff). The two most obvious forms such innovatively intensified exploitation can take are: (1) resource diversification and the more routine incorporation into the regular diet of lower ranked resources that had previously been lower ranked for their relative dietary value or because of the increased labor intensiveness inherent in their exploitation, and (2) the adoption of innovative techniques or technologies that extract more resources from the territory than did the extractive methods that they replace. More often than not, such more productive extraction methods are also more labor-intensive than those that they replace had been. This is because, logically, were these more productive methods less labor-intensive they would likely already have been employed if known (for being "easier") and population would have grown accordingly over time. If more labor-intensive methods are required to intensify exploitation in the context of a spatially reduced territory, cost-benefit calculations will likely lead to the relatively less labor-intensive ones being employed before more labor-intensive ones so long as they suffice, with increasingly labor-intensive ones coming to be employed only if/when less intensive ones no longer do so.

The point is that in the context of competition, given the ability to manage the increased risk via diversification and/or intensification, decreased mobility potentially mitigates a greater risk (the difficulties and risks of effective defense and the possible loss of critical locales) than it creates (the threat of resource depletion). The net costs versus benefits in the organic hierarchies make it selectively advantageous to those practicing it. Thus, as integral to some models, groups are not simply attracted by the lure of certain *r*-selected species in a physical environment hermitically sealed off from those groups' social environment. Rather, groups are "attracted" to such resources only in the context of competitive pressures to reduce mobility. This is because, labor intensiveness notwithstanding, such resources represent the best of the reduced exploitation alternatives inherent in the emergent system and because it is more economic (in both the perceptive and selective sense) to exploit them than to continue trying to maintain sufficient control over a larger territory.

Decreased mobility is neither a viable option nor selectively advantageous in the absence of a (reduced) territory that can be more intensively exploited than it had been or the technological ability to do so. The innovations that may permit such more intensive exploitation are simply a product of the same fundamental environmental pressures fostering compression, but in this case fostering innovation. Thus, any innovation that permits the more intensive exploitation of a reduced territory will generate territorial compression in the face of sufficiently strong competition for resources.

Intensification and the Paths to Food Production

Extractive technologies relative to available potential resources are what ultimately determine whether territorial compression can occur and the degree to which it can occur. Thus, given maximum compression for a given extractive technology relative to applicable available resources, the ability to innovate in a manner that further intensifies resource exploitation is a prerequisite for any additional compression to occur. While the competitive stresses fostering territorial compression will also foster innovations biased in the direction of intensifying exploitation, these represent independent processes.

Given territorial systems, the absence of viable conflict avoidance options, and the ability to intensify resource procurement, competition will progressively channel behaviors into engaging with whatever is the next least labor-intensive affordance available within the constraints of environmental givens, the information system, and available (or made available) technology. If the ability to profitably innovate is not available, further intensification will stall and any groups attempting to nevertheless make do with smaller territories would experience higher Malthusian penalties due to malnutrition.

Given the fact that intimate familiarity with the resource potential of their surroundings is a prerequisite for any foraging group's long-term survival, it is patently ridiculous to start from the assumption that the resources whose innovative utilization through diversification would make possible further compression were not known to such groups prior to the engagement in the behaviors that first brought those resources into a significant degree of exploitation. Likewise, it is well documented in the archaeological and ethnographic record that some of the basic knowledge concerning how to more intensively exploit a potential plant resource—for example, the knowledge that seeds or cuttings grow into plants; that small hard objects can be ground or pounded down into finer particulates—was already well known prior to when their large-scale use in the repertoire of human subsistence behaviors began. My point here is simply that human knowledge of what resources are potentially available in any given geographic area likely long precedes the ability to more efficiently exploit the full range of those resources or more importantly the willingness to do so more intensively.

It is this preexisting knowledge of what exists to potentially exploit within a territory and how to exploit that territory more intensively that makes intensification of one form or another a potential affordance, only dependent on conducive existing or new (including the introduction of domesticates from elsewhere) environmental conditions. That knowledge will preferentially channel innovative activity due to competition for resources into the domain of subsistence-related behaviors and technologies when such intensificatory affordances exist and are perceived to be more economic than alternative affordances. In essence, the widespread equifinality of some form of locally appropriate food production is due, at the proximate level, to the widely recognized fact that, as regards somatic costs/benefits, intensification when feasible is more "economic" than violence. At the ultimate level, this equifinality is due to the fact that food production allows for the maintenance of larger groups and larger groups have a competitive advantage over smaller groups as regards any potential violence that such competition can engender.

Such intensificatory innovations, in turn, may functionally permit territorial compression and make possible the evolution of new less mobile subsistence adaptations. The order of such attempted innovations will be largely based on calculations of perceived costs and benefits, meaning that, to the degree possible, less labor-intensive innovative behaviors will tend to precede more labor-intensive ones. Thus, under most circumstances where both forms of intensification are available affordances, simple resource diversification will be engaged in before production. However, both are not always possible options and sometimes neither is; one, the other, or both can cease to be an option at any point in the contingent progression of increasing intensification. This is why the process culminating in the widespread equifinality of fully settled food-producing societies is neither universal—for example, food plant production is not (yet) an option in the Artic—nor unisequential.

Take, for example, a critical resource that is perennially self-regenerating, occurs in a relatively high natural density, and is readily exploitable from relatively fixed (and thus easily defensible) locations that are also both difficult to deplete despite exploitation (by virtue of being perennial or otherwise sufficiently self-replenishing) and amenable to some degree of increasingly intensive exploitation without significantly altering the correspondence between degrees of labor input and resource output. Specific examples include nut-bearing trees and many marine resources including anadromous fish. In such cases, we can expect diversification that eventually includes such resources to then produce some high degree of sedentism centered on the most valuable extractive locales in the virtual absence of significant food production. This is because the energy expenditures needed to produce food beyond perhaps some minimal level are not required until some (often much) later time when the need for increasingly more intensive exploitation begins to strain the capacity of such perennially self-renewing resources to reliably provide for the groups exploiting them. In the absence of producible resources or unsuitable environmental conditions for production, violent competition for the locales yielding the more or less self-renewing resources will ensue when their capacity to reliably provide starts to become strained. This would, for example, tend to happen more rapidly with adaptations revolving around edible tree nuts/fruits, with their relatively inelastic maximum yields even with a degree of arboriculture, than those revolving around marine resources with their more elastic limits. We can also expect intergroup competition for such fixed locales to vary inversely with the elasticity of their yield.

An example of this trajectory would be Mesolithic northern Europe, where diversification alone ultimately allowed for a high degree of territorial compression such that the resultant adaptations can be considered essentially sedentary (Price, 1983, 1985). This abandonment of significant mobility was made possible by the incorporation of highly renewable—in this case largely marine—resources that to at least some degree could be progressively more intensively exploited in response to competitive conditions, without investing labor in more labor-intensive behaviors such as active food production. Despite evidence for violence, further intensification in the form of significant food production did not apparently occur, presumably due to the absence of native species suitably responsive on an annual basis to productive manipulation in the local environments. Significant food production did not begin there until much later, in the context of introduced domesticates and new competition from the intrusive populations that introduced them and who, by their presence as competitors, fostered larger group sizes in order to successfully compete with the newcomers. The same basic dynamic—territorial compression to the point of fully settled communities made possible almost entirely by the intensified exploitation of elastically self-renewing resources—arguably applies to other times and places where sedentary, complex hunter-gatherer societies evolved in the absence of further intensification in the form of significant food production, but the presence of violent competition. Examples of this arguably include Jomon-era Japan and the Native American societies of the American Pacific Northwest. A severely truncated version of this trajectory that stalled well before sedentism (not to mention any degree of complexity) due to inelasticity of marine resource yield (fish trapped in Nile flood "tidal pools"), if not resource yield in general, is arguably visible in the case of the Epipaleolithic Nile Valley hunter-gatherer groups inhabiting Upper Egypt.

In contrast, when diversification does not include non-depletable perennially self-renewing resources, but does result in the incorporation of resources amenable to regular propagation such as annual grasses,[6] we can expect an intermittently continuous pattern of territorial compression loosely associated with a pattern of

[6] One could theoretically argue, as do some "attraction" models, that wild cereals grasses are a perennially self-renewing resource. The problem with that argument is that, unlike marine resources and nut-bearing trees, annual grasses can be readily depleted, since the more intensively they are harvested the less will remain to naturally reseed themselves, creating a "death spiral" of availability that proceeds faster the more intensively they are exploited. In contrast, the degree to which marine resources and nut-bearing trees are exploited in any given year has significantly less impact on the next year's yield.

increasing diversification and increasing effort, with some of that effort eventually coming to be increasingly channeled into production as affordances involving further intensification via just diversification cease to become available. This increasing reliance on supplemental production allows for the continuing ability to derive sufficient resources from territories that are increasingly reduced in size as the cost of further intensification comes progressively to be outweighed by the cost of defending a larger territory.

This kind of pattern is clearly visible in the Levant and upper Euphrates drainage, where we have increasing diversification and increasingly larger number of archaeologically defined cultural entities during the Epipaleolithic (e.g., Henry, 1989). There, diversification reached its capacity to accommodate further reductions of mobility toward the end of the Epipaleolithic. However, diversification did result in the incorporation within the regularly exploited resource base of a number of annual plant species amenable to being reproductively manipulated in order to generate higher outputs through both technological (i.e., productive) and biological (i.e., selective) behaviors. The incorporation of exploitable stands of wild annual grasses in the Levant thus engendered the readily available affordance of production as a means of intensification in the face of continued competition, which allowed for a relatively uninterrupted sequence of intensification and concomitant territorial compression to the point of full-scale sedentism, such that violent competition, though very likely present from the very beginning (e.g., see Bar-Yosef & Belfer-Cohen, 2010; Grossman, 2010; LeBlanc, 2010) was apparently minimal until well into the aceramic Neolithic (see Rosenberg, 2003). However, in at least some parts of southwestern Asia outside of the Levant and upper Euphrates drainage (e.g., the Tigris drainage), there are multiple early settled village sites that did not exploit cereal grasses to any significant degree, relying heavily instead on nuts, which have fixed limits to natural yield and are difficult to produce owing to the multiyear lag between planting and harvest. It is noteworthy that at these sites we have evidence for at least some violent competition in the form of ubiquitous mace heads, at a minimum suggestive of threat displays (e.g., Özkaya & Coşkun, 2011; Rosenberg, 2011), and blunt force skeletal injuries (e.g., Agelarakis, 1993) from even the very earliest settled village communities.

In the Levant, over time this process led to increasingly larger fully settled village communities (Rosenberg, 2003) reliant on ever-higher proportions of produced food. This feedback loop came to a halt when the evolution of the full complex of regional plant and animal domesticates made possible behavioral economic complexes tailored to intensification in a wider spectrum of local conditions not conducive to further intensification before. These new economic affordances also afforded the opportunity for conflict avoidance behaviors to come back to the fore for a while as responses to intensifying competition, producing the spread of fully developed food-producing societies into at least some adjacent regions (e.g., the Mesopotamian floodplain, the Balkans), if not also the Nile delta, Caucasus, and the Kopet Dag region,[7] as well as the spread of transhumant behavioral complexes into others.

In contrast to the above pattern, a relatively low natural density of dispersed food resource locales makes significant reductions in mobility centered on the most valuable locales not a practical option even with supplemental production, which is hampered by the natural conditions producing the low natural density. In such cases, production can come into play and not lead to a significant increases in sedentism, with territorial compression and significant increases in sedentism only occurring later when more advanced productive technologies or more productive variants of those resources make their appearance and tilt the cost-benefit calculation in favor of increased sedentism by making individual locales more potentially productive. Central Mexico and the Basketmaker sequence of the American southwest represent variants of this trajectory, differing from the one in the Levant primarily in the degree of reduced mobility permitted by each intensificatory innovation due to environmental conditions. While the progression of intensification was essentially the same, with diversification leading to the regular exploitation of relatively malleable plant resources, followed by production of those plant species, full-scale sedentism was not possible in these latter cases until relatively late in the process. That is, even the inception of food production, while sufficient to mitigate the latent effects of increasing competition, was insufficient for the evolution of fully sedentary communities, which apparently had to await the (co)evolution of plant crops sufficiently productive for those environments to support further compression and fully sedentary communities.

None of the various progressions from fully mobile foraging to fully sedentary food production are an inevitable outcome of competition, even of intense competition. If critical resources are spatially restricted in their location, not amenable to any form of more intensified exploitation with existing extractive technologies, and no new extractive technologies are forthcoming, then no affordances exist to counter the stresses produced by the competition for these limited resources and Malthusian penalties inevitably come into play. A fairly well-documented example of this would be the terminal Pleistocene groups inhabiting territories along the upper Nile valley. Here, *r*-selected resource diversification was possible but beyond that further intensification was apparently not, presumably due to the characteristics of the resources being exploited and the sharply circumscribed environmental context for their exploitation. Here, the progression of intensification and reduced mobility became stalled at maximum possible diversification and maximum possible natural yield. The result was the forced maintenance of some degree of mobility, rigorous territorial defense involving a degree of preemptive offense, and, as evidenced by sites such as Jebel Sahaba (Wendorf, 1968), a level of interpersonal violence that can accurately be described as territorial warfare (Kelly, 2005). Food production, along

[7]For the moment, it remains unclear whether the latter three involve colonization or just diffusion of an economic complex.

with full-scale sedentism, did not come to be practiced there until the much later introduction of domesticates from southwestern Asia.

Also, none of the progressions of increasing effort offsetting increasingly decreased mobility are a smoothly unbroken progression. Leaving aside any subsequent higher level cultural reorganizations[8] for being too variable to broadly generalize about (due to the variability inherent in infrastructural opportunism), an emphasis on the importance of generalized reciprocity is a consistent element of the Bauplan of mobile forager group-level information systems (e.g., Boehm, 1999). In contrast, any degree of food production requires ownership at some entity level to assure both a return on the (increased) invested labor and more importantly create the ability to hold back, not share, and thus preserve working capital (e.g., seed and breeding stock) from one year to the next. In Rakudu's case (see section "The Death of Cultural Systems"), the latent hostility generated by his actions (see also Yellen, 1990) successfully pressured him into abandoning his innovative subsistence behaviors. But, as suggested in Chap. 6, if an increasing number of individuals opt to persist in such innovative contra-cultural behaviors because the benefits of those behaviors in the face of competition outweigh the social costs, it would cause the collapse of the old system's Bauplan and its punctuational replacement by a new information system revolving around a Bauplan capable of accommodating the new infrastructural behaviors and their social consequences (e.g., significantly increased endogamy at the local group level). This would allow for a meaningful reliance on produced food to supplement foraged resources, and such a trajectory is clearly visible in southwestern Asia (see Rosenberg, 1998).

Territoriality may be an innate tendency in the face of competition, but it requires sufficient predictability in the availability of whatever the nascent territory is a spatial proxy for to be worth the cost of defense and sufficient density of whatever resources the territory is a proxy for to provide for the somatic well-being of the group laying claim to it. Increased sedentism in the face of continuing competition requires the ability to more intensively exploit a spatially reduced territory. This can be done through diversification of exploited resources and the inclusion of lower ranked resources or by increased effort directed at procurement, processing, or both. Food production is simply a step along the path of increased procurement effort, but it requires the availability within the territory (or territorial suitability for introduction into it) of a resource that is sufficiently tractable that its density within the territory can be manipulated through labor input. How tractable it is determines how much resource density can be increased and how much territorial compression can occur.

The convergent evolution of evermore intensive systems of food production tends to occur wherever and whenever it is possible for two intertwined reasons. First, evermore intensive food production supports potentially ever-larger group sizes by mitigating intragroup tensions arising from competition for hitherto limited

[8] Such as the shift from round to rectilinear houses and the structural changes in the social units of production and consumption that are said to be indicative of (Flannery, 1972).

resources and thereby contributing to the maintenance of group cohesion. Second, large group size is important because it confers an advantage in intergroup competition—the advantage of greater numbers. Larger group sizes, however, require more complex forms of organization to harmonize, integrate, and coordinate the larger numbers of constituent individuals in those groups. However, the proximate processes integral to the evolution of such complex forms of organization are not as straightforward as those integral to the evolution of food production and settled village life, in that they relate less to economically coping with competition and more with actively engaging in it socially.

Chapter 9
Proximate Causation and Pattern Part 2: The Paths to Social Complexity

> *"People take the longest possible paths, digress to numerous dead ends, and make all kinds of mistakes. Then historians come along and write summaries of this messy, nonlinear process and make it appear like a simple, straight line."*
> *Dean Kamen*

Egalitarianism and Hierarchy Briefly Examined

Complex societies are now recognized to be highly variable, yet we still tend to casually think of societies as being either egalitarian or hierarchical, due to it being such a crucial social distinction because it affects a group's membership's degree of personal autonomy, among other things. But, that distinction is not as straightforward as it sounds. As pointed out by Flanagan (1989), even societies that we tend to think of as "truly" egalitarian very often have hierarchical distinctions as regards age and/or gender and even more wide-ranging formal social distinctions can exist in societies that remain nevertheless egalitarian to a significant degree. For example, material inequalities are often enough also present in societies that do not have social inequalities beyond perhaps age and/or gender social distinctions. We tend to refer to these as "nominally egalitarian" and they are egalitarian for all intents and purposes because an egalitarian ideology (i.e., information system) constrains the behavioral expression of these material inequalities as it does even in societies with the aforementioned more wide-ranging formal social inequalities.

Egalitarianism is just a set of core beliefs about the structure of equality—some defined set of individuals are equal in some particular social sense; hierarchy is a set of core beliefs about the opposite—the structure of inequality. The point is that not all social and material inequalities are equally intolerable in the context of a given Bauplan having "egalitarianism" as a central element. This is because egalitarianism is not a monolithic conceptual construct. Rather, it is an often diffuse distillation of a variable mix of concepts revolving around fairness, social equality, material equality, power over others, communal identity, etc., not all of which are of necessity elements in any given Bauplan, though the change from presence to absence (or vice versa) of any given element will typically result in a new Bauplan at some

M. Rosenberg, *The Dynamics of Cultural Evolution*, Studies in Human Ecology and Adaptation 12, https://doi.org/10.1007/978-3-031-04863-0_9

(materialistic) level. As with any other aspect of the information system, it can be understood variably by entities at all levels of the (evolutionary) hierarchies, and these variable understandings are variably expressed behaviorally and organizationally by entities in the other three hierarchies.

That means that individual systems can be more or less egalitarian depending on the mix of concepts they individually incorporate, what those concepts are understood to mean behaviorally and organizationally, and what they permit/discourage in terms of behaviors via the cultural selection sorts to the other hierarchies. Thus, egalitarianism cannot simply be said to be either present or absent. Rather, we are dealing with degrees and their expression in the variable form of egalitarianism's componential concepts and their material manifestations. The point here is that individual component concepts can be abandoned, while the overarching ethos can still remain "egalitarian" to at least some degree. The same applies to its antithesis, commonly glossed as hierarchy. In general terms, it also applies to other elements of a system's Bauplan.

The immediate point is that, as discussed in more detail elsewhere (Rosenberg & Rocek, 2019), societies are not one or the other: either egalitarian or hierarchical. Rather, we are dealing with degrees and the critical factors that make a society more egalitarian than not in any meaningful materialistic (i.e., behavioral) sense, is not the absence of material inequality or social hierarchy, but rather the presence of some mix of elements of an egalitarian ethos within the cultural Bauplan that prevents or at least inhibits some combination of the blatant exploitation of a group's individual members by other individuals, the blatant accumulation of wealth, the blatant flaunting of status, etc. Likewise, among the critical factors that may make a society more hierarchical than not is arguably the absence of such ideological constraints on aggrandizement, overt power over others, the exploitation of others for personal benefit, etc.

Not All Motivations Are Equally Motivating

As with the origins of food production and settled village life, until the last third of the last century the evolution of complex societies was viewed teleologically as the march of progress driven by technological advancements and their social consequences (e.g., Childe, 1936). Also as with the origins of food production, with the shift to causal explanations, functional adaptation became the most commonly proposed cause and complex societies were most commonly said to have evolved in order to organize and administer, because that is what highly complex sociopolitical (i.e., state) systems, of the kind most directly observable, most obviously function to do. Cultural evolution was thus again still viewed largely as a process of purposive transformation, but now rooted in the collective need to get certain proposed-to-be crucial things purposively done, be it water management, trade, resource redistribution, the conduct of warfare, or any one of a number of other proposed public goods either singly or in systemic combination. Alternately, it was proposed that complex societies evolved through warfare and/or conquest or through processes wherein manipulation of the socioeconomic system made possible the transformation of wealth (i.e., surpluses) into directive power or the transformation of limited investitures of directive power into evermore far-reaching coercive power. The evolutionary models that followed these various causal models also did not deviate from their basic approach and once again focused on selection (i.e., ultimate causation), most commonly for traits promoting cooperation.

While adaptive models for the origins of complex societies are again all implicitly agential, they here again merged proximate and ultimate causes much as they did regarding the origins of food production. The particularly critical flaw with so doing in this case is that the functions that they projected back as cause are socioeconomic and sociopolitical functions relatively important in the *historical* past. These functions are obviously important functions in multifunctioning, "fully" hierarchical stratified societies such as have existed for approximately the past 5000–6000 years. But, the possibility that the proposed functions are merely exaptations that were latent in earlier social adaptations, which first evolved to fulfil other critical functions, was never considered. These earlier functions are still fulfilled by hierarchically organized complex societies, but are overshadowed in the eyes of scholars interested in the evolution of complex societies by the socioeconomic and/or sociopolitical importance of the organizational functions in evolved, multifunctioning, "fully" hierarchical societies.

By way of analogy, what such models did was the explanatory equivalent of focusing on the child care function of school systems instead of the cultural transmission (i.e., education) function to explain their initial appearance. Child care was latent in how the school systems were structured and has become crucially important to the economic functioning of industrialized economic systems, as we recently learned with the school shutdowns of the Covid-19 pandemic, but school systems did not evolve *to* provide child care; their manifest function was to educate. People simply began to utilize the child care function inherent in the organization of school systems to, for example, facilitate the generation of two income streams within nuclear family units in industrialized societies.

The same can arguably be said concerning the organizational and administrative socioeconomic and sociopolitical functions of complex forms of social organization. This is because the advancement of a given public good, unless existentially critical is arguably insufficient grounds for individuals in an acephalous, essentially egalitarian society, to willingly cede their personal autonomy to leaders and with it open themselves up to exploitation, a danger that they are all too well aware of (see Boehm, 1993, 1999, 2000). The point is that none of the proposed public goods are plausibly so threatening, other than theoretically those revolving around warfare. Moreover, there is no need to invoke directorial power to accomplish the types of suggested public goods at the scale likely to be required at the societal scales that characterize the earliest stages of the process; one has only to look at the aceramic Neolithic site of Göbekli Tepe (e.g., Schmidt, 2012) or the early mound-building Native American societies of the American Midwest (e.g., Byers, 2011) to see the scale of public goods essentially acephalous societies are capable of effecting.

As they did with their explanations for the origins of food production, social models once again treated the problem—in this case the evolution of extrafamilial/kin group directive power in small-scale societies—as if cultural systems are completely open-ended and devoid of constraints on behavior, in particular manipulative or illegitimately coercive behaviors that expose others to exploitation. They are structured as if the group's membership is unaware of its general sociopolitical interests and will not act to protect them by forestalling encroachment on those interests. They thus fail to address the critical question of how would-be leaders could get away with engaging in typically proscribed behaviors under normal circumstances and yet still achieve their desired ends (i.e., increased power). At best, attempts to impose one's will on others in an (even nominally) egalitarian system will lead to a loss of respect, one of the primary bases for power in such systems; at worst, it will lead to the violent removal of "upstarts" (Boehm, 1999), as even Julius Caesar learned to his dismay at the hands of his nominal patrician social peers. The same applies to attempts to expand a hypothetical necessity-based investiture of limited and generally ineffective authority, as in warfare, beyond its limits. In either case, the effect is the opposite of the desired effect.

The point is that social models revolving around economic coercion or other forms of social manipulation simply do not lay out a valid culturally sanctioned path from point A to point B, the former being social egalitarianism and the latter being individual directive power over others. Economic obligation is not authority, nor can one be transformed into the other without agreement by both parties. One party can always refuse to fulfil an economic or social obligation and suffer the damage to the relationship(s) that results from the refusal. In a social system, structured and regulated by whatever degree of egalitarian ethos, attempts to unilaterally overstep the social limits of the system will be met by at least latent hostility, as with the "who died and left you in charge?" colloquial response an American reader would likely have to a fellow member of one of their peer groups who tried to impose their will on them, no matter how many gifts/favors are owed. Of course, people will tend to compete for standing to one degree or another based on personal temperament, but they must do it within the framework of the culturally structured (essentially egalitarian) social system they live within to actually gain in that competition. As noted previously, one cannot just start to tackle opposing players in a soccer game and expect to remain in the game, much less succeed in it.

Social models revolving specifically around leadership in warfare or subjugation due to warfare suffer from different problems. The problem with proposing that directive power first evolved to effectively wage warfare is that, to begin with, there is no evidence that one leads to the other despite scattered historical instances wherein it potentially should have if such models are correct. For example, in the late eighteenth through early nineteenth century, American settlers were encroaching on the territory of the Shawnee confederacy in the Ohio River drainage and a state of near-continuous war existed for almost 60 years from before the American Revolution through the War of 1812. That "endemic warfare enhanced the prestige of war leaders" relative to "village chiefs" because "the normally temporary authority of war leaders assumed permanent status" by reason that the context for handing it back (i.e., peace) never arose (Calloway, 1992: 45, see also 2007; Clark, 1993). Yet, while war leaders are said to have become more important, as is to be expected in a situation wherein their leadership became particularly relevant, they did not achieve directive power as a result; they merely retained the very limited power they typically had, and that power was not directive in the sense of what we usually consider authority. Moreover, war leaders could be deposed, as was Kishanosity in 1774. On the other hand, in cases where we do see an emergent degree of military authority (e.g., the Germanic tribes of the Roman era), we are already dealing with societies comprising numerous thousands of people (i.e., able to field many thousands of warriors) and a degree of hierarchical/heterarchical organization, as should not be surprising in societies of such size. Leaders, whether or not they have directive power, are expected to lead and that includes in warfare if that is included in their charter, but that does not mean that directive leadership evolved in order to conduct warfare. Models that rely on warfare to elevate leaders arguably presuppose a Bauplan that already legitimizes directive power.

The problem with proposing that exploitative hierarchical social structures evolved through subjugation as a product of warfare is different. It is that such a model requires a system of extrafamilial/kin group hierarchy and exploitation to already be in place before such a model becomes operational. In other words, the starting conditions any such models arguably require are the conditions whose evolution they are trying to explain. Models that rely on subjugation presuppose some preexisting internal degree of involuntary appropriation such that the idea of absorbing others in order to systematically appropriate their labor and/or resources is already a valid conceptual option. More importantly, they further presuppose that social features and ideological foundations are already in place to make such appropriation both possible *and worthwhile*. Were either/both of these not already present, the winners would simply kill the losers or drive them from their territory and appropriate their territory for themselves and their descendants (e.g., as described by Rappaport, 1968), because that would arguably be the optimal benefit of victory under a *preexisting* egalitarian form of organization, both in terms of how the winning group would perceive benefit and in terms of objective contribution to the winning group's biologic fitness.

Lastly, as with the beginnings of food production, evolutionary models tend to here focus exclusively on ultimate causation, more often than not specifically on selection for cooperative behaviors in large-scale groups (e.g., Richerson & Boyd, 1999). True, centralized, directive leadership is more efficient than acephalous organization at organizing cooperative endeavors and thus advantageous in intergroup competition, but focusing solely on selective processes ignores the cultural constraints on behavioral selection that under normal circumstances would prevent elements integral to the evolution of directive leadership from being practiced to be selected for and no proximate processes capable of precipitating the elimination of those constraints (e.g., through systemic dis-integration) are typically offered. Such elements are just implicitly assumed to be available to the operation of selective forces. Moreover, efficient cooperation is all selectively well and good. But, complex societies typically also employ at least some degree of involuntary appropriation that arguably acts to the somatic detriment of the vast majority of the group and the evolution of that contradiction also is never addressed.

Leaving aside the details of the many models that fall under these various headings, the point once again is that many make valid points. Thus, they are not necessarily wrong in any absolute sense, they are at best again simply incomplete for omitting proximate causation. There is, however, more often than not a second problem. It is that they are also often too linear; the evolution of sociopolitical and socioeconomic complexity is significantly more variably complex than the evolution of food production. In the same manner of speaking as used before, we are here not even speaking of "elegant semicircular pegs," but rather variously sized pie slices that can be variably assembled along with the minor sector(s) representing proximate forces to fill the "circular hole" that is a coherent explanation (see Rosenberg & Rocek, 2019). This is because even though competition is again central to proximate causation, it drives proximate causation in a number of different ways, more often than not only indirectly, and producing multiple possible outcomes that do not lead to convergence in anything other than in thus far unlimited maximum possible group size. This is because the proximate causes underlying the evolution of food production involved economically coping with competition, often at an infrastructural level, while the evolution of complex societies was and still is a product of actively engaging in competition opportunistically at a structural level.

The Preconditions for Social Complexity

While degree of social complexity is intertwined with group size, which is ultimately contingent on food production and the ever-larger groups that food production is theoretically capable of supporting, the evolution of sociocultural complexity produces a much more varied range of trajectories than does the evolution of food production. This is because competition proximally plays out more directly as a social process (as opposed to a technological one) than it does for the evolution of food production. This means that the evolution of social complexity is driven much more by calculations of social costs and benefits, as opposed to directly somatic ones, than is the evolution of food production. That is, there are a wider variety of options available for effectively harmonizing, integrating, and coordinating individuals than there are for keeping them sufficiently fed, and while it can theoretically be argued that some options are more efficacious than others and will therefore be selected over time as a consequence of intergroup competition, that does not mean they all have not been tried or that the specific differences even matter in a given competitive context.

Forms of social organization and social structure more complex than those that characterize mobile foragers are all contingent on the existence of surpluses, whether produced or naturally available, that can be directly or indirectly used to mobilize individuals to collective action and to support group members not directly involved in the production of those surpluses; the larger the surpluses the greater the potential for such complexity. Surpluses, whether of naturally occurring or produced resources, are a logical outgrowth of any subsistence system that requires the time-limited collecting/harvesting of resources, as it would be recognizably foolish to stop such accumulation at estimated normal requirements and not allow for unforeseen events that would have to be met after the resources required to meet them are no longer available. Such unforeseen events, being uncommon—for otherwise they would be included in the calculation of "normal requirements"—mean that more often than not the extra resources collected to be available to meet them will eventually turn out to be true annual surpluses.

Surpluses are owned by some level of organic entity and not automatically subject to expected sharing with other organic entities of the same or higher order. They are thus incompatible with the generalized economic reciprocity integral to a typical mobile foraging cultural system, for otherwise they would be immediately distributed and not be true surpluses. As noted previously, the production of surpluses thus requires a new type of cultural Bauplan, as would arise out of the systemic disintegration of a Bauplan having generalized economic reciprocity as an integral ideal.

Any new cultural Bauplan underlying surplus-producing subsistence systems must accommodate ownership of economic property at some entity level in order to preserve working stock and thus it incidentally permits the differential accumulation of wealth by entities and the development of material inequalities, though it may also promote the redistribution of such accumulations and thus minimize (but not eliminate) such material inequalities. However, material inequalities are not the same as formal social inequalities (beyond those normally rooted in family and basic kinship obligations) of the sort usually referred to as hierarchical. Though often treated as if they proceed in tandem, social and material inequalities do not necessarily go hand in hand. The former without the latter constitutes a form of social relations usually referred to as nominally egalitarian and it is possible to have degrees of hierarchy in the absence of *significant* material inequalities, as in the Pueblo tribes of the American southwest or the aceramic Neolithic societies of southwestern Asia (see Rosenberg & Rocek, 2019). Social inequalities (i.e., hierarchy) invariably confer authority on dominants over subordinates and require deference to dominants by those subordinates; that is all they invariably do, though the temptation by dominants to use them to self-interestedly further their material well-being is invariably latent and will sooner or later become manifest unless assiduously checked culturally.

The point is that looking to material inequalities as the fulcrum for the evolution of sociopolitical inequalities in human societies glosses over a critical question. It does not address the crucial question of why egalitarianism is abandoned, because it represents the starting condition and is highly valued by its practitioners. Egalitarianism, at least within any age and gender social divisions, is a tenet in the Bauplan of virtually all cultural systems associated with mobile foraging societies and generally labeled as egalitarian. It is thus the baseline state from which more complex sociopolitical systems evolved. Being highly valued, it is not lightly abandoned and one cannot get to the evolution of hierarchical relations associated with sociopolitical complexity unless it is first abandoned.

Social animals, humans included, by nature tend to be hierarchical because hierarchy has selective value at both the individual and group levels. This is because, at the level of the individual, higher standing yields both somatic and reproductive benefits and, at the group level, hierarchy prevents most conflicts of interest within the group from escalating into violent confrontations that jeopardize the cohesion of the group. It does so because individuals know their approximate current relative standing in the group and, therefore, whether or not they are likely enough to win a direct confrontation to make the social and/or physical risk of such a confrontation worthwhile.

Egalitarianism is a cultural construct that has evolved in anatomically modern humans that counteracts this natural tendency to hierarchy. Just as the tendency to hierarchy is intertwined with the natural tendency to act self-interestedly, so is egalitarianism rooted in that same tendency to act self-interestedly, coupled with the tendency to act cooperatively. As detailed by Boehm (1999), an egalitarian ethos is firmly rooted in recognizable self-interest for the large majority of individuals in a given group. This is because for the majority of a group's members, the odds of being at the top of a given hierarchy are recognizably slim. So, for that majority, it is better that there be no meaningful hierarchy at all. In essence, it is an informal coalition of the group's majority—the latent subordinates in any potential hierarchy—that aims to prevent the latent dominant(s) from ever achieving that position and so to forestall the latent threat of domination/exploitation that hierarchy represents. Egalitarianism is therefore fiercely defended by a variety of behavioral mechanisms, including ridiculing pretensions, ignoring demands, rewarding modesty, generalized reciprocity (preventing the accumulation of obligations), etc., and if all else fails the violent elimination of "upstarts." The proper question is thus not: how did hierarchy evolve in human societies? It is: if human beings value egalitarianism so highly—and they do, as witnessed for example by its central place in virtually every major western utopian philosophy to be proposed since at least the enlightenment—what prompted human groups to abandon egalitarianism in the first place?

Theories for the origins of social complexity have traditionally focused on the evolution of those complex societies characterized principally by authority at the societal level, surplus appropriation by hierarchically defined social elites, and ostentation by those elites. All three are a product of competition and all three require breaks from a Bauplan having egalitarianism as a central element, but not all require the same breaks, nor are any of the three contingent on others of the three, meaning they evolve independently of each other, not necessarily in any specific order, and it is possible to get the evolution of any one alone, or a combination of any two, as well as of all three (see also Rosenberg & Rocek, 2019).

Hierarchical authority obviously requires a break from the most basic element of an egalitarian ethos: equal social standing, and while both appropriation and ostentation appear to commonly co-occur with hierarchy, they are not theoretically contingent on hierarchy. That is, both the latter two can be by an acephalous group as a communal whole and while calling the necessary contributions in such cases always "voluntary" is perhaps stretching that term's meaning a bit, the point is that the impetus to conform by contributing in such cases is peer pressure and not directive power.

Ostentation, being inherently wasteful of typically material resources, is contingent on a Bauplan that includes the toleration of owned surpluses that can then be wasted, typically by some subset of a community, and the abandonment of generalized reciprocity at the community level. It is not contingent on either hierarchy or appropriation. Owned surpluses act to undercut egalitarianism by removing one of the mechanisms promoting it (generalized reciprocity), but the toleration of owned surpluses does not necessarily go hand in hand with the wholesale abandonment of egalitarianism and it certainly does not require hierarchy. That is, it remains possible to tolerate economic ownership and still pay lip service to egalitarianism in nominal form, as when you still have an egalitarian ethos in the absence of strictly egalitarian behavioral practice, exemplified by differences in accumulated surpluses but the absence of individual ostentation (see Flanagan, 1989; Rosenberg & Rocek, 2019; see also Spikins, 2008). However, it is possible to have ostentation and an egalitarian ethos coexisting if the ostentation is by the group as a whole, in which case nominal egalitarianism can persist until egalitarianism is fully abandoned in connection with the toleration of individual ostentation, the latter also requiring the abandonment of a communal ethos. If and when that happens, individual ostentation can then either evolve on its own or coevolve with hierarchy when that also evolves, if it has not already done so.

Appropriation is also contingent on the toleration of owned surpluses, whether of resources or time and energy. It is not contingent on ostentation at all and only certain forms are contingent on hierarchy. If it is by the group as a whole, even if it seems to benefit only a limited number of members, then appropriation can coexist with an egalitarian ethos. On the other hand, appropriation by some subset of the group (whether or not that subset claims to be doing it for the benefit of the group) is contingent on sociopolitical structures able to compel compliance. Such structures may be hierarchical, but they can also be rooted in fear of supernatural sanctions, such as those at the disposal of ceremonial sodalities.

The Ostensible Benefits of Abandoning Egalitarianism

Adaptive/functional explanations for the evolution of complexity tend to explicitly or implicitly revolve around the relative organizational efficiency of centralized decision-making processes over group decision-making processes—the more centralized the better, because the political feasibility of agreement on a decision is progressively less of an issue the smaller the number of people involved in making that decision.[1] This makes swifter group-level responses to events possible. In essence, such explanations propose that the group's membership will, of necessity or choice, abandon egalitarianism in order to organize some public good whose benefits outweigh the loss of personal autonomy and the concomitant latent threat of exploitation for the majority destined to be formally subordinated. The basic problem with such explanations is that the ostensible benefits proposed by these models are only marginally beneficial at best for the majority of the group and thus highly unlikely to be seen by them as worth the latent risk of exploitation. In other words, not only is the sort from the information hierarchy normally working to oppose innovation in the form of hierarchy, so are the sorts from the two organic hierarchies.

To begin with, there is ample archaeological evidence that many if not most of these functions can be implemented sufficiently well (in the context of the sociodemographic starting conditions) by acephalous societies and even by relatively small kin groups within larger societies. For example, the earliest aceramic Neolithic societies of southwestern Asia, for which there is currently still no evidence of directive leadership, were already able to construct the megalithic enclosures at Göbekli and engage in the long distant trade of obsidian. Even *organized* trade could be conducted by sub-societal groups within later complex societies, as with the early second millennium BC kin-based Assyrian trading networks attested to by the tablets from the Karum at Kanesh. As for organizing warfare, the aforementioned Shawnee, whose war leaders did not have real directive power, were doing sufficiently well in their conflict with American colonists until 1783, when the British withdrew their support as arms suppliers (Calloway, 1992: 47) in the context of peace negotiations with their former colonies. The point is that one can easily argue that extrafamilial/kin group hierarchical organization is unnecessary for organizing the proposed public goods at the sociodemographic scales that constitute the societal starting conditions from which hierarchical organization is proposed to evolve.

Moreover, aside from directive power arguably being unnecessary as regards organizational ability at the earliest stages of the process, a good case can be made that emergent elites derived significantly greater net benefits from any and all such organizational services than do the large majority of subordinated group members. For example, large-scale irrigation/terracing/flood control/etc. systems provided larger surpluses for elites to appropriate. They likely did not provide significantly increased surpluses for subordinates to retain or use, because those would by definition be *surpluses* and thus available to inevitably be appropriated for use by elites. Likewise, religious activities and edifices validated the emergent elites' parasitic statuses and provide the subordinated majority with no more supernatural aid than they also provided the elites. Similarly, militaries protect the large majority of members from external threats no more than such militaries also protected the elites themselves, while they also make possible the projection of the elites' power

---

[1]This says nothing about the good judgement inherent in one form over the other, just the relative efficiency of the decision-making process.

(i.e., the ability to coercively appropriate) over the group's membership as well as over "others."

The case for similarly skewed benefits is arguably also possible for the functional benefits from any other such public services. For example, *organized* trade was typically at least somewhat weighted toward luxury goods and did not benefit the majority at all except to the degree it fed into one of the above functions (e.g., scarce construction materials for public edifices).[2] Even direct redistributions of critical resources in times of crisis would arguably be beneficial to the large majority of members only because the productive group's (e.g., household's, local community group's) surpluses, which would otherwise have been retained by such groups for their own emergency use, had already been largely, if not totally appropriated by elites. Thus, such potential surpluses were unavailable for independent storage and subsequent independent use by households or local community groups in such times of need. In a very real sense, such direct redistributions of resources only serve to minimally mitigate the parasitism of elites, when necessary, by maintaining their subordinates at some minimal level for survival. Thus, it serves the interests of elites as much as those of the majority to maintain their subjects through such direct redistributions when the need arises. This is because, bluntly put, a parasite's fitness suffers if it lets its host die.

The root of this problem is that the benefits that are proposed are organizational functions performed by full-scale complex societies of the type typically called states, the need to fulfill which functions is then viewed as the proximate cause for the evolution of such societies. However, since such functions are not truly beneficial to any meaningful degree to the large majority, they arguably only evolved after hierarchy had already been established for some other purposes that the majority viewed as sufficiently beneficial to them. That is, the commonly proposed explanatory functions are arguably exaptations built on directive structures which evolved to fulfill earlier functions that were the basis for the abandonment of social egalitarianism in the first place. They were arguably not the basis for the abandonment of egalitarianism themselves because they are arguably not seen as sufficiently beneficial to the majority who would be required to give up their personal autonomy to gain the purported benefit.

In the absence of the majority's willingness to accede to a would-be leadership's attempt to make and impose decisions on the group, any attempt by would-be leaders to do so would at best lead to a loss to them of the only form of then-existing political power—influence—for being seen as too power hungry, and at worst lead to their elimination for being too serious a threat to egalitarianism within the group. This eliminates bullying, attempting to overtly obligate others by using wealth or any other existing socioeconomic currency, or expanding any necessity-based limited mandate (e.g., leadership in war) to other areas of life not within that limited

[2] If absolutely essential resources were in such short supply as to warrant their importation, then it is unlikely that the region in question could have supported groups of sufficient size to drive the evolution of sociopolitical complexity in the first place.

mandate. Thus, we must look to a general, ongoing function that benefits the group's membership in truly equal fashion and that also immediately outweighs the cost of latent and, given the temptation to engage in self-interested behaviors, ultimately inevitable exploitation of the to-be subordinated majority by dominants. The key once again is the behavioral tendency to compete.

However, unlike the case with the origins of food production, in this case we are dealing with two concurrent levels of competition. On the one hand, we are again dealing with intergroup territorial competition, which continues unabated, differing from expression to expression only in whether it drives violence or further territorial compression and packing through more intensified exploitation of resources to the degree possible. On the other hand, we are also dealing with the normal latent interpersonal competition within groups, only now fueled by differential surpluses and so no longer merely latent. The key is the threat such intragroup competition poses to cooperation within the group and thus the group's ability to effectively compete with other groups.

Competition, Group Cohesion, and the Evolution of Directive Power

Human beings are arguably the most complexly cooperative social animals on the planet and the most intensely group selected at all social levels. As individuals we have been selected to seek the safety and security of a social group and we strive to be accepted by the groups we belong to or seek to belong to. Were it otherwise, peer pressure would be ineffective and it is anything but that. Even "nonconformists" conform to the norms of whatever nonconforming group they belong to, outlaws conform to the norms of whatever outlaw groups they belong to, etc.; hermits are rare outliers.

The one consistent feature of social complexity—and complexity is not here being used as synonymous with hierarchy—is that it roughly correlates with group size; the greater the complexity of its social structure and organization, the larger the group such complexity is capable of organizing and holding together on a permanent basis. In competitions between the two, larger groups generally tend to be at an advantage over smaller groups and, size aside, groups with structures that allow for more centralized decision-making tend to have an advantage over similar-sized groups with more decentralized decision-making, in that they can mobilize to collective action more nimbly and effectively. Leaving aside the issue of which drives which—it is at least partly circular, the ultimate function of social complexity is maintaining group integrity and the ultimate cause for the evolution of complexity is that it was selected for because it allows for the maintenance of group integrity, and hence cooperation, by increasingly larger groups (e.g., Richerson & Boyd, 1999); and, all else being equal larger groups tend to be more successful than smaller groups in intergroup competition.

The basic social problem here is that the number of potential dyadic pairings within a group—the elemental social level at which conflicts of interest can materialize—grow much faster than the number of individuals in the group.[3] Conflicts of interest at some level of seriousness are inevitable in even the strongest dyadic relationships and differ meaningfully only in the degree of threat they pose to the relationship. Thus, in the absence of effective mechanisms to resolve conflicts of interest within a group, the larger the group the more likely it is to eventually fission due to serious conflicts of interest arising between individuals in relatively weak dyadic relationships and the faster that eventuality will materialize. In addition, the larger the group and hence number of relatively weak dyadic relationships within it, the less effective peer pressure would be at averting any potential tragedy of the commons and the group-wide conflicts of interest integral to one, not to mention the threat the tragedy itself would pose to the group as a whole.

As a top-down group selected social species, we naturally desire social harmony within the group—doubly so when confronted with an external threat, because our individual welfare is strongly tied to that of the group. Conflict resolution is an important aspect of a dominant's role in our nearest primate relatives, the (common) chimpanzee (see De Waal, 1982, 1996), who also engage in violent intergroup competition, and it continues to be an important function of leadership in human groups whatever their organization (e.g., See Boehm, 1999, 2000; Richerson & Boyd, 1999), from the very simplest, up to and including the most ephemeral human forms of hierarchical leadership (e.g., see Maybury-Lewis, 1974: 203–204), right through to the largest state systems wherein the ruler(s) or their delegated proxies (i.e., courts) adjudicate disputes.

Human intergroup competition can vary from latent at one end of the spectrum to extremely violent at the other, based on the demographic ceiling imposed by any given extractive technology in a given environmental context. The more violent is such competition, the greater the premium on relatively large group size because there is strength and safety in numbers. Thus, level of intergroup competition is a factor fostering increased group size, while social structure and organization effectively caps it. In an essentially egalitarian system, if intergroup competition is relatively low, the latent threat to a group and its constituent individuals posed by fissioning is relatively low and there will tend to be less peer pressure exerted by the rest of the group on the parties to a dispute to accept a mediated resolution. This makes such fissioning more likely to occur than if the threat posed by potential fissioning were higher and the group's membership therefore inclined to exert more forceful pressure on the parties to resolve the dispute. Thus, the more intense is intergroup competition, the more likely are groups to be larger because relatively large group size tends to be increasingly seen by the membership as an asset in that competition and the membership will tend to be inclined to exert stronger pressure in an effort to get the parties to accept a mediated resolution and thereby maintain group integrity.

---

[3] $C(n,2) = n!/(2!(n-2)!)$

However, there is a limit to the effectiveness of such informal conflict-resolving mechanisms' ability to successfully resolve conflicts of interest, be that by simple peer pressure or supernaturally buttressed pressure by, for example, shamans or ceremonial sodalities. This is because, in the end, the willingness to accept a mediated resolution is strictly voluntary and thus dependent on a host of factors, not least of which is the parties' history of past conflicts of interest. Whatever the level of pressure may be to accept a resolution to a dispute, one or both parties to that dispute can always simply refuse to and the group will consequently fission, which under conditions of violent competition is a potentially existential threat to both the splinter and rump groups and their memberships. It is under such an existential threat that the ability to impose a resolution on the parties becomes invaluable for maintaining group integrity and it is under such conditions of *current* violent competition that the investiture of some level of blanket social authority to ensure social peace within the group arguably comes to be seen by the membership as outweighing the risks to the majority of potential *future* exploitation. Recent historical examples of such subordination of individual autonomy, in these cases by group-level entities, to a new higher order entity in order to maintain/create internal harmony in the face of external threat can be seen in the formation of the League of the Ho-de-no-sau-nee by several tribal-cum-societal entities collectively usually called the Iroquois (see e.g., Morgan, 1851) and the nucleus of the European Union by several nations (i.e., societies) beginning in 1952.[4] While these examples involve the relinquishment of individual autonomy by groups to the leadership of a group of groups, there is nothing to indicate that the maintenance of internal peace within their group is not of equal concern to the individual members of groups under external threat.

Thus, it is the internal peacekeeping function of leaders in times of external threat, not their value as military leaders or their value as organizers that arguably serves as the initial catalyst for the investiture of authority in them. Once fully established by a group to maintain group cohesion, other functions can come to be assumed by such directive leadership as exaptations because they further enhance the group's ability to compete with other groups through greater nimbleness of decision-making and/or organizational efficiency. Moreover, there is potentially another equally direct benefit to the individual members of the group in their entirety from such an investiture of the power to resolve conflicts and maintain cohesion: the ability to exploit other groups. It will be explored below.

An important point requires clarification here. Violent competition will not inevitably lead to the investiture of overt directive power in leaders. Alternative trajectories that accomplish the same social ends (i.e., conflict resolution to maintain cohesion) within the group do exist. Of these, the most obvious is the social control

[4] In the context of the escalating cold war, these first steps toward "European federation" through the "pooling of coal and steel production" under "one high authority" were designed to make (internal) war "not only unthinkable but materially impossible" (Schuman, 1950).

role[5] often played by ceremonial sodalities, such as those so common among the Native American tribes of the American southwest. If they already exist for other social and/or ceremonial reasons, then the social control function can easily evolve as exaptations of those institutions (see Rosenberg & Rocek, 2019) and separate investitures of directive power in order to maintain social cohesion are not required to allow group size to continue growing.

In the absence of preexisting alternative structures (e.g., ceremonial sodalities) that can be coopted to take on a sufficiently powerful conflict resolution function as an exaptation, it is only in the context of violent competition that the group's majority would see it to be to their *net* current advantage to subordinate their personal autonomy to leaders with directive power. The threat of exploitation no doubt remains an obvious concern, but it is not a current threat, while fissioning is a current threat. Thus, to invoke the distinction apparently made by some Native American tribes (e.g., see Lowie, 1973), it is one of the primus inter pares leader of the "peace chief" or "civil chief" type[6] who is likely to be voluntarily granted directive power to speak for the group on the basis of collective self-interest, not the "war chief" type who somehow transforms his limited (and consciously curtailed) military power into more generally applicable power through coercion or manipulation. This is because the former would seem the least likely to abuse such power to exploit subordinates. Would-be dominants need not manipulate the system in the modes typically invoked by social models to gain directive power against the group's natural resistance to granting it; they are voluntarily vested with it by the group as needed to maintain cohesion and thereby any form of relative group advantage (see Rosenberg, 2009). They only need to maneuver through mollifying expressions of generosity and fairness (e.g., see Stanish & Haley, 2005; Spikins, 2008) *within* the preexisting egalitarian constraints of the social system to be the individual(s) so selected.

While the granting of directive power to a leader will typically result in at least the partial disintegration of a social system having an egalitarian ethos at its core, it is also possible for some elements of an egalitarian ethos to be carried over into the new system and remain in force to dampen any behavioral expressions (e.g., ostentation, exploitation) of any such new hierarchical structures.[7] However, if that investiture is then expanded to include appropriation for purposes of (other than group level) ostentatious displays, the result will be the disintegration of what had already become only a semi-egalitarian system and the abandonment of whatever remained of an egalitarian ethos.

---

[5] As noted by Brandt (1994:17), "There is... fear of persons...that possess [the] power... to control fundamental forces...: the person who can cure can also harm; the [sodality] that can bring rain can also cause a flood...."

[6] This applies whether or not this is a formal distinction for the group in question.

[7] This is because there can be a difference between cultural ideology and cultural practices—i.e., ideal and real (see Flanagan, 1989; Rosenberg & Rocek, 2019).

Competition and the Evolution of Exploitation

While directive power opens the door for exploitation on the part of a leader that wields it, a Bauplan that allows for the former does not automatically also allow for the latter. One can easily argue that exploitation arises independently of directive power (but potentially in parallel with it) and in a much more organic fashion in the same context as that which promotes the investiture of authority on leaders: violent competition.

The suggested role of violent competition in the formation of complex societies was most systematically presented by Carneiro (1970, 1998). Simply summarized, he proposed that in highly circumscribed physical environments, winning groups would absorb the losers prevented from leaving by such circumscriptions, thus establishing the exploitive social hierarchies that characterize complex societies. However, I propose to turn that around. I suggest that the ultimate roots of social complexity lie not in the emergent relationship of the losers to the winners in some violent competition, but in the relationship of the losers to the relatively friendly groups with whom they are forced to seek refuge when driven off their territory.

Aside from the basic problems with warfare and subjugation models detailed above, there is another, in this case conceptual problem with the mechanics of Carneiro's model. It is that no matter how tightly circumscribed any given area may be physically, the individual local groups within it are nevertheless invariably still surrounded by multiple neighboring local groups in a dynamic social environment. That is, it is difficult to envision a circumscribed physical environment so small as to be populated by just two small (by virtue of starting conditions) local groups—any physical environment so small would simply be too unstable to support a demographically viable human population on a long-term basis.

Such other neighboring groups may be either related at some level or not, and friendly, neutral, or hostile. While any conflict between such groups could potentially involve large shifting alliances made up of multiple such local groups, ultimate defeat would still tend to befall a very limited number of local groups at a time (e.g., see Rappaport, 1968) because egalitarian societies have neither the numbers nor organization to displace more than one or at most a very few local groups at a time. The point is that even within a highly circumscribed physical environment, losing local groups would still have other friendly, allied, and/or related local or even more distant groups with whom to take refuge, even in the highly unlikely event (see above) that the winners would somehow wish to absorb the losers as a group in order to exploit them.

While it is possible to define contexts wherein such refugee groups would not be at a social disadvantage, as would be the case if their presence was seen by the hosting group as serving some useful purpose (e.g., establishing them as a buffer group), the weaker the historic relationship between the two groups the more likely that the refugee group would in fact be at a latent disadvantage, however subtly, if for no other reason than they would now be in a dependent relationship to the hosting group.[8] In any such dependent situation, a refugee group would be highly susceptible to subtle manipulation, if not subtle coercion, by their hosts to their disadvantage and the relative economic and reproductive advantage of the hosting group, the more so to the degree that the refugee group could be seen as some degree of "other" and thus potentially exempt from even just nominally egalitarian behavioral obligations.

The point is that, given normal human tendencies to pursue self-interested goals to the degree socially permitted (by one's own group), any refugee group of "others" in such a position can easily be so manipulated to at least some minimal extent. More importantly, any relative advantage held by a hosting group depends on the maintenance of cohesion within that group, in that any conflicts of interest producing a split would potentially negate their social advantage. At worst, it could result in a realignment that produces two new groups, both having drawn members from both the refugee and hosting groups, neither of which would now have a social advantage. At best, it would water down their social advantage by reducing the advantaged group's size and concomitant ability to manipulate the refugee group to their advantage.

Spencer (1993: 48) suggested that authority initially comes to be invested in leaders when the "success of individuals [becomes] dependent on the success of the group." I suggest that it is precisely in such subtly group advantaged contexts that this occurs. It is a social given that, even while part of a social group, individuals will pursue their own interests and those interests will at times be in conflict with those of other individuals, if not those of the group itself. Such conflicts, if not resolved, have the potential to splinter the group; and, for members of an advantaged group, such a fracturing would weaken their position of relative advantage, thus jeopardizing the welfare of the group as a whole and most if not all of its individual members. I suggest that being in such an advantaged context is yet a further powerful impetus for the advantaged group to formally delegate directive power to one of their own in order to facilitate conflict resolution in order to maintain their perceived individual *net* advantage relative to the individuals of the hosted group even after their subordination to the new, formal dominant is factored into the equation.

In other words, I suggest that it is the same inherited human tendency to form alliances (see De Waal, 1982, 1996), which Boehm (1999, 2000) suggests underlies an egalitarian society's ability to maintain its egalitarianism that also underlies emergent social hierarchies and exploitation. In the case of egalitarianism, it is a coalition of the latent "underdogs" opposing the latent "top dogs" that serves to maintain that egalitarianism and forestall potential disadvantage to the coalition's members. In the case of emergent hierarchy and/or exploitation, it is a coalition of the same latent "underdogs" *with* a latent "top dog" that serves to ultimately grant the

[8]The relationship of the Shawnee to the various groups they lived among in their movements through the eastern United States over the course of the seventeenth through eighteenth centuries, in response to pressure by both the Iroquois and European settlers, serve as examples of the full range of such potential relationships (see Clark, 1993:62ff.).

latter formal dominance and so maintain the group's collective at-this-stage-still informal dominant position as a whole.

However, *any form* of structured inequality is a radical departure from any form of preexisting egalitarian system. The resultant conflicts and contradictions with any remaining egalitarian elements of the Bauplan other than communalism will result in a rapid disintegration and reorganization of the entire sociopolitical system (see Rosenberg, 1994) around a new Bauplan incorporating a broadly and generally applied system of formal social hierarchy incorporating directive power. Such rankings will more likely than not be structured around kinship and supernatural descent, simply because kinship is a structural element that can readily be carried over without conflict from the older, abandoned system. In any case, once any form of inequality is structurally sanctioned, even if only internally, it has the potential to be extrapolated to any and all other applicable circumstances (e.g., intergroup relations), and the dominance of one group over the other(s) inevitably becomes formalized as well. This includes formal dominance over any hosted groups.

I further suggest that at this early stage of the process, hierarchically organized complex societies typically do not yet grow by the forced incorporation of other groups in the fashion Carneiro proposed. Rather, I suggest they more often than not grow by continuing to dispossess other groups of their territory and occupying any such newly acquired land themselves through population growth (i.e., the latent tendency for population to grow, made manifest by the availability of additional territory to grow into) and/or attraction of friendly and likely related groups that voluntarily join them in the context of the social arms race that has now begun. At this stage, the level of resource appropriation engaged in by the dominant subgroup is likely to still be relatively mild and advertisement in the form of group-level aggrandizing behavior for purposes of attracting members may evolve as a feature, a point I will return to below. Lower ranked groups are not likely to vigorously contest the emerging new social order to the degree that, given their only limited "otherness," they likely identify at some significant level with the dominant group, and their subordinate status is only marginally so.

However, resource appropriation, once institutionalized at even some low internal level is now, like social inequality, also culturally available conceptually for extrapolation to social relations marked by even greater degrees of conceived "otherness." Moreover, the temptation to thus appropriate (i.e., get something for "nothing") from groups where the scale would be unchecked by the constraints of group identification can be assumed to be a given. It is at this stage that Carneiro's model becomes operational.

Authority-based centralized leadership, once established, enhances organization and that affects group-level success in competition with other groups. But, potential organizational improvements are not the precipitant of the social innovation, the need to maintain group cohesion is. Once directive power is established in some organic group in the economic hierarchy,[9] at a minimum group size will remain stable due to effective conflict resolution, and can now even continue to grow. Thus, there would rarely be a context where the basis for that investiture will expire before it becomes well established as a long-term cultural norm. Once established as an accepted norm, it is capable of even being carried into new environments by relatively small breakaway groups, who may not yet need the basic public good of *large* group size maintenance (e.g., see Kirch, 1984).

Competition, Display, and Ostentation

As argued elsewhere (Rosenberg & Rocek, 2019) and also by others working from a variety of perspectives (e.g., Leonard & Jones, 1987; Yoffee, 2005), the evolution of social complexity is not linear and complex societies do not readily conform to types. The evolution of formalized directive power is proximally driven by the need for the maintenance of group cohesion in the face of intergroup competition, but the evolution of directive power is an entirely different matter than the evolution of the organizational structure of such power and also of the trappings of such power. Complex societies vary in the historical order of their constituent institutions' development, the rate of their individual institutions' development, and the specific forms such developing institutions take. All that matters is that emergent institutions are congruent with the then-existing Bauplan, or failing that, the crystallization of a new one around these new institutions.

While the evolution of directive power may be proximally processual, the subsequent evolution of directive structures is shaped primarily by engagement with both physical and social environmental affordances: the interaction of opportunity and ability, which can vary widely. A trading empire, such as that controlled by the Medieval Venetian Republic, will be organized differently than a military/tributary empire, such as those controlled by the Aztec or Inca. Thus, while the form any initial investiture of political authority takes may typically have a common basis, the form of its organization is unlikely to. The organization of directive power is obviously contingent of the prior evolution of such power. However, once directive power becomes established, the organizational structure and functions of directive leadership evolve as opportunistic exaptations to accommodate emergent organizational needs, as enhanced organization makes possible new collective endeavors made desirable by environmental factors and/or historical opportunity. It does not follow some predetermined developmental trajectory. The evolving organization of power is a product of the group's organizational needs, itself a product of what new collective endeavors need to be administered (e.g., social harmony, warfare, taxes, tribute, trade).

---

[9] It almost always exists in the genealogical hierarchy even in "truly egalitarian" societies (see Flanagan, 1989).

However, the organization of power is different than, and independent of its trappings, such as dramatic and ostentatious displays of culturally sanctioned leadership statuses, as well as displays of any coercive power associated with such statuses and their ideological foundations.

The evolution of such trappings is proximally rooted in behaviors that typically predate the evolution of both directive power and appropriation, though they do require surpluses.

For social animals display can serve both intra- and intergroup competitive purposes. Intragroup displays serve to enhance individual standing relative to potential rivals and so deter rivals from mounting overt challenges, as well as serving to attract mates. Intergroup, group-staged displays (e.g., howler monkey and gibbon choruses) serve to deter rival groups from mounting territorial incursions, as well as perhaps intimidating rival groups to more readily submit to incursions. For humans, they do essentially the same, and given the complex sociality on display in the structure of the economic hierarchy, also serve to attract members to the group. Thus, the trappings of power are very much a product of what they are intended to signal to potential allies or rivals, who those potential rivals are, and whether they are rival individuals within the group or rival groups. In other words, the trappings of power are all various forms of display.

Displays need not be ostentatious, aggrandizing, or even involve material goods at all. As simply behavioral phenomena, they can predate the evolution of economic systems that produce surpluses. For example, as in other species, martial displays as preludes to fights staged on "fight grounds" (e.g., see Rappaport, 1968: 121ff.) as well as any other collective action intended to advertise group size and perhaps aggressiveness are designed to intimidate rival groups by simple reference to available brute force. But, ostentatious and/or aggrandizing displays require surpluses.

In human societies structured by the framework of a culture that has an egalitarian ethos as a central tenet of its Bauplan, intragroup competition is purposefully dampened in order to maintain such egalitarianism behaviorally and that includes suppressing overt displays of "difference" meant to enhance an individual or subgroup's standing. Ostentation, if attempted, by individuals will thus have the opposite of its typically intended effect and at a minimum generate ridicule in place of respect. This applies even in surplus-producing nominally egalitarian societies, wherein such surpluses do foster interpersonal competition, but wherein that competition is still structured along a surplus-accommodating, modified version of egalitarian principles, such as spending or giving away wealth for the ostensible purpose of the group's benefit.

As long as both an egalitarian and communal ethos are in force, no matter whether behavioral patterns fully conform to those ideological ideals or not (by pushing limits), ostentation by individuals is frowned upon. In such cases, even if formal leadership statuses do exist in the context of a new, residual "egalitarian" system that accommodates them, we still have what Spikins refers to as "prestigious leadership" and not "coercive leadership." The distinction between the two being that a "prestigious leader or authority will be careful not to claim to be different to or better than others" (2008: 179), while the latter does so claim. In other words, some modicum of modesty (i.e., paying at least lip service to an egalitarian ethos) is a feature of the former manifestation of hierarchy, but not the latter. Prestigious (i.e., consultative) leaders give voice to the group; coercive (directive) leaders speak for the group.

Prestigious leadership does not allow for aggrandizing behaviors by individuals; directive leadership does allow for them, and such behaviors by individuals evolve as an expression of the intragroup competition for directive leadership statuses. The emergence of aggrandizing displays as an expression of blatant interpersonal competition for status, in turn, is a potential outgrowth of group-level ostentatious display. The election node determining allow/deny individual aggrandizement is only contingent on the availability of surpluses and the preexistence of group-level aggrandizing displays.

Such group-level displays are an expression of intergroup competition and such group-level displays can predate the evolution of directive power. Intergroup competition is constant and ongoing, whether for territory and its resources by societies or, as in modern times, for profits (e.g., corporations), jobs (e.g., cities), membership (e.g., sodalities), students (e.g., universities), etc. It can vary over time in the degree of competition and by type of group, but it is always at a minimum latent for groups of the same kind. That competition can also differ in its expression, such as being channeled into rule-limited violence (e.g., competitive sports) for competing subgroups of a larger group that must still maintain at least nominally amicable relations in order for the larger group to maintain cohesion. It can also be channeled into display.

All individuals are essentially ethnocentric to some degree. That is, if a group's members do not believe that the information system shaping the group's organization and behavior is not more correct than (i.e., superior to) any alternative systems available for adoption as systems,[10] the members would tend to drift away and adopt the "superior" system of whatever rival group they believed shared and lived by such a "superior" system.[11] By way of example, if I, as an American, truly believed that the British, Japanese, or Mexican way of life (as a system) was superior to the American way of life, I would be sorely tempted to pack up and move to Great Britain, Japan, or Mexico as the case may be, and become British, Japanese, or Mexican. The (voluntary) immigrant ancestors of most Americans largely did just that, not to mention the more current waves of immigrants doing the same. The same applies if I came to believe that the information system of any of my sub-societal (e.g., religious, political, etc.) groups was inferior to any of the available alternative systems; if so I would likely "convert."

It therefore behooves groups to materially and if appropriate ostentatiously display their superiority over rival groups, as for example through edifices (as proxies for wealth, cohesion, etc.), iconography, and displays of armed power. Superiority can also be displayed through competitive redistributive exchanges and

[10] As systems, not as unintegrated assemblages of traits.

[11] "Superior" does not imply perfection, just better than its comparator.

the implicit one-upmanship inherent in the gift itself: we will see your previous gift of 400 pigs and we will raise you[12] by adding a further 200 pigs, 12 cassowaries, A $10,000, a pickup truck, and a motorcycle—top that if you can. The purpose of the displays is both to deter rival groups and reinforce the membership's sense of group identity and ethnocentrism, and thereby reinforce group cohesion, by valorizing the group. The implicit message to other groups is: we are so much more powerful, legitimate, supernaturally favored, prosperous, energetic, industrious, cooperative, and/or well-led than you that an overt challenge is not worth mounting. The implicit message to the group's membership as well as potential allies and/or recruits is: you cannot do better than us; the grass is nowhere greener.

Such group-level aggrandizing behaviors do not automatically produce or even automatically allow for individual aggrandizement. The election node governing the evolution of individual aggrandizement may be contingent on the preexistence of group-level aggrandizing displays and intragroup competition for leadership statuses, but it is not determined by them. Instead individual aggrandizement is arguably contingent on the leadership status coming to symbolize the group itself and a leader becoming an integral element of the group's aggrandizing displays. More widespread individual aggrandizement follows as an expression of the competition as displays of worthiness regarding leadership status.

The elevation of individuals to be symbolic of the group is the key to the evolution of aggrandizing behaviors on the part of individuals. The relationship of that symbolization and individual aggrandizement is still visible in modern complex societies at both the societal and even sub-societal group levels. For example, the American president—the head of state, chosen collectively by the group, lives in a mansion, has a private chef, a private plane, a private helicopter, a private retreat, and a host of other perquisites. As a group, Americans believe their economic system, for example, is the most powerful and efficient[13] presently known and America's wealth and productive capacity are held up as proof of that. If true, do Americans really want their leader living in a corrugated tin shack, hitchhiking to meetings, and eating off of paper plates when hosting foreign leaders, since that would visibly belie that proof? As a group, Americans do not begrudge appropriation in the form of taxes for that purpose, but will typically begrudge the leader's "retinue" (the broader leadership and lower level governmental workers) those same "perks" because they do not symbolize the group at all. One can see a more extreme form of the same distinction in the peasant uprisings of medieval Europe (e.g., Jacquerie), which were primarily directed at the rapacious nobility, but not the king, whom the rebels truly (but erroneously) expected to side with them and attend to their grievances (Tuchman, 1978).

---

[12] This is a poker-parlance reference to the Kawelka's Moka gift in the 1970 documentary film *Ongka's Big Moka* (also known by the title: *The Kawelka*), directed by Charlie Nairn. The fact that poker often involves bluff is integral to the analogy.

[13] This is not a statement about fairness or anything other than productivity and rank efficiency.

However, the evolution of such symbolization is independent of group-level aggrandizement and the competitions driving it, in that it can remain absent even in relatively complex societies, such as it apparently was in the urban societies of Indus civilization, which apparently maintained a communal identity even after the abandonment of the other elements of egalitarianism within that cultural system's Bauplan (see Maisels, 1999). The symbolization of leaders is likely most often rooted in kinship seniority, since a weaker form of such symbolization is arguably inherent in genealogical entities. The point is the evolution of such symbolization is contingent on an election node that is not biased by the competitions driving group-level aggrandizement. However, once symbolization is established, individual aggrandizement becomes driven by an extension of that same process in the form of individual worthiness.

The message embedded in group displays also varies based on its purpose. As long as a complex society's membership remains restricted to a single, however, loosely self-defined and self-identifying (tribal) group and possible real or fictive cadet branches, the legitimacy of dominants is implicit in the information system shared by the group. Leaders symbolize the group and followers identify with leaders to some degree as embodiments of the group itself (e.g., see Levine, 1967). It is in this kind of social context that the evolutionary mechanisms explored by Richerson and Boyd (1999) operate most effectively to produce complexity. As long as warfare by such a group is geared to competing with other groups for members and space into which to expand, highly visible propagandistic displays ostentatiously legitimating the power of dominants, as opposed to ones reinforcing group identity and cohesion, are only minimally necessary. However, once—for whatever reason—the goal of warfare ceases to be the displacement of other groups for purposes of demographically fueled expansion, and instead becomes the subjugation of groups sufficiently "other," for the explicit purpose of exploiting them in one capacity or another, then more prominent and dramatic representations of the validating ideology and the symbols of its power become necessary. They do so in order to more effectively control groups who can be made to bow to, but nevertheless might otherwise vigorously contest the legitimacy of the dominants, because they do not naturally identify with the dominants at the relevant social level or subscribe to its information system.

The shift in goals from displacement to exploitation is also not inevitable. The reasons for such a shift in goals no doubt vary, but more often than not likely revolve around simple historical opportunity. That is, appropriation for the benefit of the group is already a feature of social relations by virtue of a preexisting Bauplan that accommodates aggrandizing group-level displays, which by their very nature require appropriation of surpluses by the group. Thus, opportunities to more ruthlessly appropriate from groups deemed sufficiently "other" will be seized upon when they present themselves. This would particularly be the case if the territory occupied by such "others" cannot be as effectively utilized by the winners through demographic expansion (e.g., due to distance or spatial discontinuity) as it can be by leaving the losers in place and exploiting them. It is when such opportunities to do arise and are taken, that Carneiro's (1970) mechanisms for more marked class formation, and Spencer's (1998, 2009) for greater decentralization and specialization come into play and setting up the potential for still greater complexity and even larger social entities.

The Evolution of Social Complexity Is Complex

As should be obvious by reference to just the three commonly cited characteristics of substantially complex societies considered above, the evolution of social complexity is arguably much more complex and nonlinear than we tend to appreciate. While all three of the above-discussed end states are contingent on the preexistence of a cultural system that permits economic property in the form of surpluses or its derivatives, none of these three states are contingent in their initial form upon the preexistence of any of the other two. That is, what we have are three independent elemental trajectories stemming out from a common starting state, proximally driven by engagement in different aspects of competition, meaning that there is no invariable order to their evolution. Moreover, these three are not the only traits that can enter into the evolution of complexity, nor are all three theoretically even necessary for substantial complexity.

Complexity can take the form of any combination of these and other traits. Though some traits may co-occur to make possible (but not determine) specific outcomes of particular interest, such as directive power and appropriation by the group to make possible appropriation by individuals, their individual evolution does not necessarily proceed in any specific order relative to each other. The fact that the state of C is contingent on the existence of states A and B does not mean that B is contingent on the existence of A or vice versa, and that A of necessity had to evolve before B or vice versa. It simply means that both need to exist, whatever the order of their appearance, for the election of state C to become possible. For that matter, because A and B evolved independently of each other, one could have evolved while the other never did, in which case the potential election node that could potentially result in the state of C never becomes manifest.

For example, you cannot have a system that accommodates any kind of exchanges beyond generalized reciprocity unless you first have one that already accommodates personal property. Likewise, while a system cannot accommodate the flaunting of such property unless you obviously already have one that accommodates the accumulation of said personal property, the accumulation of personal wealth is not a sufficient condition for the flaunting of wealth even though it is a necessary condition. There must also at some previous or *subsequent* point have been the abandonment of the egalitarian ethic prohibiting individual ostentation and the node that resulted in that election is independent of the one that produced then-current concepts of property. Thus, it is possible for the Bauplan structuring fully complex societies to nevertheless retain vestiges of egalitarianism, as for example does the Bauplan structuring American society, or that which structured the Medieval Republic of Venice (e.g., see Crowley, 2011).

Every social system is, at any moment in time, the current product of multiple specific trait-related allow/prohibit elected states that have often enough evolved independently of each other, even if they are contingent on the same previous state, even if they are integral to the same larger system, and even if some number of them may all be both fostered by and selected for by competition. Changes to those elected states are constrained by the framework of the larger system's conceptual Bauplan and what its structural logic deems permissible regarding socioeconomic and sociopolitical states of being, such that the evolution of foreclosed states requires the abandonment of the then existing Bauplan and the crystallization of a new system that simply permits them, regardless of what other states the new system also potentially permits or forecloses.

A system will evolve within the constraints of its Bauplan, but become increasingly rigid as it does so due to the growth of structural constraints created by previous cultural elections, not to mention the vested interests they produce. The more complex it is, the more readily this will likely happen. When it reaches the limits of its ability to adapt to changing circumstances within the confines of the Bauplan's conceptual constraints and/or the then-existing structural constraints, there are one of two outcomes. The first is that the system will become increasingly ineffective and gradually lose out in the face of competition with other groups organized differently (see Spencer, 1998). The second is that the group abandons it outright or the system disintegrates as the group's members, in attempts to more effectively compete, increasingly engage in behaviors that violate the system's conceptual constraints, causing its conceptual collapse due to a growing inability to police the increasing violations.

In either of the latter two cases, the old system will be rapidly replaced by a new system having at its core a Bauplan that accommodates the behaviors integral to the abandonment/disintegration of the old system. The new system may involve wholesale changes or it may substantially carry over the bulk of the old system and involve just sufficient changes to accommodate the precipitating behaviors. In either case, those changes may include reversions to elements abandoned at previous election nodes, as with a reversion to egalitarian elements in state systems. It could also be a reversion to a structurally simpler and therefore substantially different system. How changed the new system is affects how many prior structural constraints have now likely been removed and how many vested interests have been eliminated for being made immaterial. Accordingly, the more changed the new system, the greater the potential to eventually evolve in for-the-moment unconsidered directions completely foreclosed prior to the disintegration of the old system, though there is no certainty or necessarily even a high probability that it will do so. The changes simply create future opportunities for the system to evolve in directions previously foreclosed by creating new states that possible future states might be contingent on.

How much a particular cultural system shifts in one direction or another, at what election node—for example, permit or forbid economic property; extrafamilial hierarchy; exploitation; aggrandizement; etc., when, in what order, and in what manner depends on what affordances and which specific constraints are present or absent at that time. While competition proximally drives the behaviors that precipitate new systems, those new systems are also capable of incorporating drift-produced traits largely peripheral to the current form of organization, but essential to some future organizational outcome potentially contingent on their presence. There is not just one overarching contingent sequence that will ultimately generate complex sociopolitical systems from some starting state of "true" egalitarianism. Rather, we are dealing with a plethora of individual, often independently evolving traits and contingent sequences that combine to produce a myriad of specific systemic sociopolitical outcomes.

Chapter 10
Agency, Proximate Causation, and the Sloshing Bucket

"Too often we forget that genetic microevolution is, after all, only one specific example of evolutionary process. Upon careful inspection, we can see that cultural evolution is disanalogous in many ways to that example. Yet culture is still an evolutionary system in its own right, and a powerful one at that."—William H. Durham.

Summary

When the social sciences adopted a Darwinian evolutionary framework in the latter third of the twentieth century as the proper perspective for viewing the evolution of socially learned human behavior, the modern synthesis still held sway and that had come to be dominated by genetic reductionism. In contrast to Darwin's own thinking, biological entities above the gene were seen as ephemeral and thus borderline irrelevant to evolution; it was genes that critically competed, not individual organisms. Individual organisms were at once viewed as being mere vehicles for the differential propagation of genes and the implicitly passive subjects of environmental forces based on their genetic traits, with no consideration given to how individuals might react physiologically, behaviorally, or any other way besides genetically to the environmental stressors that drive selection.

Evolution was thus only a matter of ultimate causation, and only as so far as concerned genetic traits. Proximate causation—agency, was considered irrelevant because, Dawkins' metaphor notwithstanding, genes have no self capable of being agentially selfish. Moreover, the different evolutionary processes—selection, inheritance, development, and the production of novelties—had come to be seen as all fully independent of each other, or "fractionated," as Walsh (2015) critically terms it, such that they do not influence each other at all.

As applied to the evolution of culture, the same focus on only ultimate causation largely held sway. The fundamental problem with this almost wholesale importation of the modern synthesis into the social sciences, as pointed out by some (e.g., Boyd & Richerson, 1985; Durham, 1991), was that cultural evolution is not completely analogous to biological evolution. Specifically, humans are arguably the most

M. Rosenberg, *The Dynamics of Cultural Evolution*, Studies in Human Ecology and Adaptation 12, https://doi.org/10.1007/978-3-031-04863-0_10

consciously and complexly purposive agents in the planet's biome and to focus solely on ultimate causation implicitly treats human entities of whatever kind or level for all practical purposes as simply and passively subject to environmental forces, be they a product of the physical or social environment, which determine their relative reproductive fitness. However, human beings are anything but passive in their interactions with their social and physical environments. A proper view of cultural evolution requires a complementary focus on behavioral selection because proximate causation sees entities as purposive agents capable of behaviorally availing themselves of environmental affordances as they perceived them.

Environmental forces do not *cause* either physical or behavioral novelties. Leaving aside the issue of whether and how environmental forces can *e*ffect physical change in an organism (e.g., see Walsh, 2015 for a discussion) for being tangential, they can generate desires in organisms capable of having them and however vaguely or sharply the organism's mind is capable of framing them. Those desires precipitate purposive behaviors, which may or may not be novel. Whether novel or not, the competition integral to those environmental forces will then select for any physical attributes or other behaviors that best facilitate the state of being produced by those purposive behaviors. But, the state of being has to first exist for selective forces to favor it and its supporting elements. And, for a novel state of being to exist there first has to be some physiological or behavioral novelty. To ignore behavioral selection and thus agency in generating novel behaviors, and thereby novel purposive states of being, particularly in sentient species, defies both common sense and observable reality; the more sentient the species, the more so.

The genetic reductionism of the modern synthesis was eventually challenged by punctuated equilibria and its theoretical fallout, evolutionary developmental biology, niche construction, and our evolving understanding of molecular genetics, leading to the still evolving extended synthesis, which has since been filtering into our understanding of the evolution of culture. Our developing understanding of molecular genetics came to contradict the wholly gene-centric view of evolution at the heart of the modern synthesis, because genes and phenotypic traits are now known to simply not be linked in the efficient manner required by genetic reductionism, and entities above the trait have come to be increasingly recognized as being subject to selection. After all, as per Darwin, it is individual organisms that struggle for survival and to reproduce.

Equally important for present purposes, the divorcing of evolutionary processes from each other—their fractionation—and the resultant structural inability of the modern synthesis to accommodate agency can be seen as implicitly challenged by the evolutionary dynamics of Eldredge's system of hierarchies, with individual organisms at their nexus—the bottom of the metaphorical sloshing bucket (Eldredge, 2002). More directly, however, it was explicitly challenged theoretically by niche construction and philosophically by Walsh's (2015) call for a situated Darwinism, both of which revolve around at a minimum implicit agency.

Situated Darwinism, and more importantly for present purposes the agency integral to it, necessitates dealing with behaviors, be they innate, learned individualistically, or acquired socially (i.e., culturally). As regards culturally transmitted behavior, because culture is the product of a distinct evolutionary system intertwined with that which governs a species' biological evolution,[1] Eldredge's dual hierarchies approach to the problem of biological entities, taken alone, is insufficient for fully modeling the evolutionary dynamics governing the interplay of all the different entities involved in cultural evolution. A second set of hierarchies is required, based on the interplay of information/ideas and behavior integral to the mechanics of culture change (e.g., Goodenough, 1981, 1999; Boyd & Richerson, 1985), that loosely parallel Eldredge's replication and interaction hierarchies' structural function as concerns evolutionary dynamics. This produces a set of four hierarchies, two organic and two metaorganic, with the individual still the nexus of the enlarged set. Each is governed by a distinct mode of selection, with the sort from each one constantly and simultaneously flowing into each of the other three and providing varieties there that each of those hierarchies' distinct mode of selection can act on. In this expanded version of Eldredge's approach to evolutionary dynamics, the individual, only now sometimes represented by the organic individual's metaorganic proxies, remains the bottom of the sloshing bucket, with lower order entities than the individual largely veiled from *direct* selective pressures by virtue of being just embedded components of the larger complex systems that constitute individuals or their metaorganic proxies.

Agency is purposive and goal-oriented. But, that goal is not to progress or even necessarily to adapt, so it is not teleological. It is simply to achieve something, whether or not that something is to adapt to a particular set of circumstances. More importantly, the purposive orientation of agency is completely consistent with what we know about the behavior of, if not all animals, at least those with some critical degree of sentience and so humans most of all.

Ultimate causation, as concerns both biological and cultural evolution, is in the final analysis a product of selection in the organic hierarchies. Leaving aside as tangential proximate causation in biological evolution (e.g., see Walsh, 2015), for cultural evolution proximate causation is a product of the metaorganic hierarchies, most immediately behavioral selection in the adaptive hierarchy, usually based on the sorts coming into it from one or more of the other three hierarchies, but not necessarily in compliance with all (or even any) of those inputs. This is because different sorts can foster opposing goals. Individuals and groups can and often enough will engage in metaorganically promulgated behaviors that seem to jeopardize their biological interests in order to further culturally defined interests. Sometimes they actually do harm to their biological interests, the fate of the various levels of Shaker entities being an extreme example of this. Other times, the jeopardy some biological interests are put in may very well constitute a roundabout way of furthering other biological interests that are calculated to be a priority over the ones being jeopardized or, more commonly, simply provide a net overall benefit. An example of this would be altruistic behaviors that sacrifice somatic interests (e.g.,

[1] This applies to any species that can be said to have culture, no matter how primitive or even nascent it may be.

sharing food, heroism, etc.) and by doing so directly or indirectly further reproductive interests (see Zahavi & Zahavi, 1997). In any case, the selective forces in the organic hierarchies—ultimate causation—will eventually weed out the poor mathematicians and along with them their effect on the sorts flowing to the metaorganic hierarchies (and from there the sorts flowing back to the organic hierarchies).

On the other hand, individuals and groups will sometimes engage in behaviors fostered by the sorts from the organic hierarchies that seem to harm their culturally defined interests—social acceptance based on perceived degree of conformity to the ideational system that structures behavior. They do so primarily by engaging in behaviors that violate that system's structure and by doing so risk social penalties that will manifest themselves in the organic hierarchies, prompted by the sorts flowing to them from the informational hierarchy. In other words, by doing so they potentially harm their social interests in the very same organic hierarchies whose sorts precipitated the behavior(s) in the first place. In essence, they potentially harm in a different way the very same organic interests that fostered the behaviors (s) in question and which interests the behavior(s) were intended to promote. The basic calculation affecting their persistence thus boils down to whether or not the contra-cultural behavior(s) provide a greater benefit to those interests than do the cost of the social penalties they incur and thereby provide the incentive to persist in those behavior(s) despite the social penalties.

All but the highest order entities are members of next higher order entities, and all but the very lowest order entities are thus complex systems. Systems, whether ideational, behavioral, or social, have a structure, for otherwise the system would not function as a system. That structure has a structural core—the system's Bauplan—on which the system was "built" around and on which in can continue to be "built," but only in ways supported by that structural core. Moreover, systems become more rigid over time as changes to the system accrue by virtue of having been accommodated by the system's existing structure and its Bauplan. That is, changes to the system that its then-existing structure allows become features of the system's evolving structure. As changed features of the systemic environment, they, like the system's preexisting features come to be purposively used by entities for their benefit, creating ever more new vested interests that resist further changes in otherwise accommodatable (by the Bauplan) directions that might jeopardize those benefits to them.

The point is that agency is not unbridled. It is constrained by the structure of the system(s) in which context a given purposive behavior is engaged in, and most of all by the system's structural core—its Bauplan. Not all novel behaviors engaged in by entities are compatible with the next higher order system of which they are a constituent element. The degree to which a novel behavior threatens the integrity of the system's systemic structure affects the degree of resistance the system will exhibit to the novel behavior. Systems are open to potential change, in the form of novel behaviors, in directions potentially left open by the system's structural core, as modified by the further constraints of its evolving structure, closed to potential change in directions foreclosed by that system's structural core, and hampered by the system's evolving rigidity. In other words, novel behaviors are not all equally accommodatable by a given system. Novel behaviors which can be justified by conceptual extension of the system structural core can potentially be accommodated if they can overcome the system's increasing rigidity. Novel behaviors that cannot be so justified in order to be accommodated are penalized. A designated hitter to replace the pitcher when batting does not violate the basic structure of baseball and (while anathema to purists) was allowed by the system's structure, which already allowed player substitution; a soccer player picking up the ball and running to the opponent's end of the pitch does violate the game's very systemic structure and is thus foreclosed as a potential change to how the game can be played.

A system's structural core, its Bauplan, is the most resistant to novel behaviors or ideas that threaten its structural integrity, the information system's Bauplan—its eidos and ethos—most of all. This is because an information system consists of abstract ideals, including those concerning how things should properly be, whereas entities in the other hierarchies—and those in the adaptive hierarchy most of all—are required to mediate between these abstract ideals and perceived practicalities—what works best. Thus, in the face of environmental pressures in the organic hierarchies, it is quite possible for the sorts flowing from the organic hierarchies into the adaptive hierarchy to overpower the constraining sort flowing into it from the information hierarchy and produce contra-cultural innovative behaviors intended to alleviate those environmental pressures that entities will persist in despite the social sanctions from other (non-innovating) entities. Such a behavioral "insistence" on "building" onto the behavioral system in a manner not supported by that information system's structural core will result in a disintegration of both the behavioral and informational system/subsystem and its very rapid replacement by a new system capable of accommodating the innovative behaviors that precipitated the old system's collapse.

Such punctuational episodes occur when local stresses grow to the point of precipitating persistent contra-cultural behaviors sufficiently threatening to the integrity of the system or when colonization of new environments exposes the colonizers to new stresses the ancestral system is incapable of handling as well as some contra-cultural innovation can. In either case, the new system will only tend to incorporate salvageable elements of the old system to the degree they are compatible with the persistent innovative behaviors the new system must accommodate and any further innovative behaviors based on them. This systemic disintegration and the rapid crystallization of what replaces it is why the break between locally successive named archaeological entities is typically so rapid and the reorganized new system is often enough so dramatically different that archaeologists are legitimately able to debate whether those successive local entities represent a case of in-migration or the evolution of a local group's cultural trappings.

Particularly good examples of this process are visible in the origins of food production in a given area. Just to be clear, this does not refer to the gradual microevolution of food-producing economies growing ever more reliant on produced and eventually domesticated resources. Rather, it refers to the rapid and pronounced changes in a group's other cultural trappings (e.g., living arrangements, concepts of economic property, attempts to garner supernatural aid, etc.) that takes place at the very outset of this (subsequently microevolutionary) process as the old immediate return system and its attendant social relations collapses and is rapidly replaced by a delayed return system and its attendant social relations.

It is the aggregate agency of individuals that is the proximate cause of these punctuational episodes, who by persisting in actions resisted by an existing system's structure, if not foreclosed by the system's Bauplan precipitate its disintegration. Agency is purposive, but a specific agent's specific intent is never reliably knowable. Even were an individual alive to ask their intentions, there is the question of whether the reply is truthful, and even were that so, whether the individual is fully aware of their own motivations. For higher order entities intent is even murkier, given the possibility that the group's constituent entities may have different if not cross purposes (e.g., see Okasha, 2018). However, intent (and thus agency) can be surmised from patterned outcomes produced by some number of similar entities (e.g., see De Waal, 1996). Just as one person behaving oddly different is considered an eccentric, so are solitary higher order entities; and just as a pattern of largely similar behavior by some significant number of individuals constitutes a purposive norm, so does the pattern exhibited by some number of higher order entities, with the patterned outcome indicative of the purpose. For entities above the level of the individual, that purpose is the mediated/additive/subtractive sum of constituent lower order entity purposes—the sloshing bucket at work once again. In that sense, such higher order entities are not strictly speaking agents, but can nevertheless be thought of in agential terms in the same manner as genes can be thought of as selfish despite being incapable of selfishness (see Okasha, 2018).

Darwinian evolution is ultimately a product of selective forces, and selective processes always require competition between competing alternatives, for otherwise there would be no force to select from among them. Drift may accrue in an entity, but sooner or later that entity will be in competition with other entities in which such specific drift-produced changes did not accrue, making the drift-changed entity subject to selective processes, with reference to which drift-produced changes the selective processes will also operate. Competition between like and/or unlike entities is thus always integral to Darwinian evolution and human organic and metaorganic entities are no exception to this evolutionary axiom.

Evolution is an historical process and thus subject to chance at the level of both proximate and ultimate causation. But, as regards behavior, proximate causation is a product of agency, which is biased by any innate behavioral tendencies characteristic of that biological taxon, making particular responses to particular environmental conditions more likely than others and amplifying the patterning produced by a given set of conditions. In other words, such patterning will be both most pronounced and thus best understood as regards intent when it is produced by innate behavioral tendencies. Such tendencies include, but are by no means restricted to territoriality and display as responses to competition. These tendencies are most directly expressed in the sorts from the organic hierarchies to the adaptive hierarchy and produce behavioral biases in the latter hierarchy.

Competition for resources commonly produces territoriality in environmental contexts wherein territory is a reliable proxy for resources. Territoriality typically maximizes day-to-day reliability of access to the territory's resources, minimizes day-to-day competition for those resources, and confers defensive advantages should the competition for those resources require active defense. However, territorial defense carries costs, not the least of which is the risk of injury, if not lethal injury, meaning that the benefits of defending a given territory have to outweigh the potential costs. Given more intense competition, the need to actively defend a given territory instead of simply relying on display will increase, increasing the risks of injury and thus the costs of defense.

If an organic entity utilizes an affordance to more intensively utilize its territory in novel ways, it can potentially reduce the difficulty and possibly also the costs of defense by allowing it to reduce the size of its territory while maintaining the territory's overall sufficiency of yield. Leaving to philosophers, biologists, and psychologists the issue of where in the continuum of organismic awareness of self and surroundings the fuzzy boundary between pure chance and true agency lies, the greater that awareness, the more readily it will purposively do so. It is in that process of assessing costs and benefits regarding territorial defense as opposed to intensified resource exploitation in a competitive environment that the proximate causes for the beginnings of human food production can be found.

That competition will also independently tend to foster behavioral changes, biased by both innate behavioral tendencies and also by sentient problem-solving to favor novel ones that make potential affordances more readily engageable, select from among the behaviors so fostered, and typically favor those that provide the greatest benefit for the lowest cost. An affordance, once engaged with, will tend to constrain what further affordances potentially become readily engageable, as well as constrain the behavioral options for readily engaging with them, and those behavioral options will again be biased by any innate behavioral tendencies. The variability in outcomes as concerns intensification is largely the product of what affordances are or become available.

Reacting to competition for resources by way of technical/technological behavioral innovations that permit abandoning elements of that competition and subsisting on a limited by nature and now spatially reduced physical environment typically offers fewer behavioral options than does proactively engaging in such competitions with other entities. This makes the latter and the social complexity it can engender substantially more variable in both its purposes and the specifics of its manifestations.

Competition, while not necessarily continuing at a constant level, is at a minimum latent if not ongoing, and engaged in by organic entities at all levels, minimally with like order entities. All else being equal, larger groups will tend to be at an advantage over smaller groups in that competition, if for no other reason than by virtue of the sheer force of numbers and doubly so if that competition is violent. For social groups, social cohesion is paramount, particularly in the context of violent competition, because the greater the cohesion, the more willingly individuals behave altruistically, the greater and more effective is thus the possible group-level cooperation, thereby making manifest the full latent advantage of larger size, and thereby the more successful the group will tend to be.

However, the larger the group, the more difficult it is to maintain group cohesion in the face of the increasing likelihood of serious conflicts of interest arising. In other words, the larger the group the greater are the centripetal forces needed to offset the centrifugal forces such conflicts of interest represent. Maintaining social peace is a typical responsibility of human leaders in groups of all sizes and levels of complexity, as well as dominants in at least our nearest primate relatives. It is this need to resolve conflicts of interest within the group and so maintain social harmony, particularly in the context of violent competition, that is the basis for the investiture of authority in human leaders. Once extrafamilial directive leadership is elected, a new sociopolitical system crystalizes around it.

Maintaining internal social peace is arguably the most primitive function of directive leadership and the basis for its evolution. This is because it is the only function that provides a perceptible existential benefit, in the form of maintaining group cohesion in the face of violent competition, that outweighs the loss of individual autonomy to dominants and the concomitant latent risk of exploitation by those dominants. Once established for that purpose, as needs or opportunities arise, organizational functions evolve as exaptations made possible by the preexistence of directive leadership and as directive leadership is seen to be capable of directing them when previous more egalitarian organizational configurations were not, or capable of organizing them on larger scales and/or more efficiently than previously.

Actively competitive behaviors can range from displays, including those that implicitly threaten violence, to outright violence. Display is both less risky than violence and, when the competition is for resources, typically less expensive than abandoning the competition and permanently losing access to the resources being competed for. Thus, displays typically constitute the inceptive form of an active and overt competition. If successful, displays can potentially deter overt challenges by intimidating competitors, thus avoiding potential violence. They can also potentially impress possible allies, thus potentially increasing the chance of success, by virtue of having allies, should the competition become violent.

Displays can be performed by both individuals and groups. But, overt displays by individuals conflict with the modesty and communalism of the most comprehensively egalitarian systems, because they make manifest latent intragroup competition, thus jeopardizing cooperation and group cohesion. For that reason, they tend to be frowned upon in such systems. On the other hand, displays by the group do not so conflict and also serve useful functions, making overt group displays arguably the older and more primitive in any given type of group. In addition to messaging directed at potential competitors and allies, group displays can also enhance the group's sense of identity and thus cohesion by valorizing the group. The form and messaging inherent in the display depends on whether it is primarily directed internally, externally, or both and, if external, whether it is directed at potential enemies, allies, or both.

Overt displays valorize the entity performing them, ostentation is a potential element of such displays, and the more ostentatious the display, the more it does so. Thus, ostentatious displays on the part of individuals are particularly discouraged in all but the most minimally egalitarian systems. Individual ostentation, initially on the part of leaders, evolves when the leader comes to symbolize the group. At this stage, it only minimally conflicts with an ethos of modesty and communalism because the leader's ostentation can be seen as valorizing the group symbolized by the leader and not the leader as an individual. However, once established, more widespread individual ostentation can then evolve in the context of overt contra-cultural behavioral competition for leadership statuses. If it does, it will result in the crystallization of a new accommodating Bauplan that has been stripped of any emphasis on individual modesty.

Exploitation as an aspect of social complexity, as first suggested by Carneiro, evolves in the context of group-level violent competition, but arguably not in the exact manner he proposed. Instead of resulting in the suggested manner from the winners in a violent competition absorbing the losers in order to exploit them, it arguably first arises from the social disadvantage the losers are to the members of the friendly/allied host group(s) they take refuge with and on whose goodwill they are thus dependent to at least some extent. This allows the hosting groups to subtly, if not more overtly, pressure the refugees into acting to the advantage of the hosts at the cost of their relative disadvantage (e.g., giving up women in marriage). How quickly such oppositional behaviors become formally legitimized hinges at least in part on the duration of the dependent relationship, the ability of the hosting group to maintain cohesion and thereby their advantage, and the degree of "otherness" that can be attributed to the refugees and thereby the degree to which any group-limited egalitarian obligations can be waived with respect to them. Once even such subtle exploitation becomes established as a behavioral pattern, whether or not it has yet become culturally acceptable, the concept of exploiting "others" becomes available for purposeful application to evermore clearly defined others. But, engagement with that affordance, and the imposed oppositions such engagement produces, only becomes possible if the organizational structures to effectively exploit such clearly defined others become available for that specific or any other purpose and that is contingent on hierarchy and directive leadership being in or coming into place.

Lastly, it is a given that any particular process transpires in the context of environmental conditions, such conditions including any other evolutionary processes as might also be transpiring. But, the fact that the transpiration of one process might, reflexively or not, be impacting the transpiration of another one does not alter the fact that they are separate processes, with such impacts only affecting the specific dynamic form of the outcome(s).

Some Final Thoughts

The shortcoming of explanations invoking only ultimate causation's role in the evolution of culture is that they boil down to explanations that explicitly state only that something evolved because it was selected for, ideally also accompanied by a statement of why it was selected for. The problems lie in the accompaniment, itself necessary to raise the explanation above a mere statement of the obvious.

To begin with, there is the potential pitfall of adaptationism (see Gould & Lewontin, 1979). Particularly for humans, metaorganically promulgated, but organically maladaptive behavioral systems can persist for a very long time before selection in the organic hierarchies weeds them out, and it may never even do so if the behavioral systems are deeply embedded in and thus obscured by larger behavioral systems that generate net benefits. Evolutionary Psychology and Human Behavioral Ecology, for example, tend to avoid that trap by often invoking evolved behavioral tendencies, in turn explained by the independently evolved human capacity to calculate economically. They thereby implicitly invoke proximate causation to explain why the behaviors were/are engaged in to be selected for. However, neither addresses the evolution of cultural systems.

Proper explanations for the evolution of cultural systems need to do the same and in the same manner, because the bigger problem with invoking just ultimate causation to explain cultural evolution is that it omits plausible explanations for why given novel behaviors were first engaged in, when one can often enough easily argue that the behaviors are proscribed by the evolutionary progenitor of the system that contained them. Selection can only operate on what already exists. Novel behaviors are not usually produced mechanistically in the manner of mutations. Therefore, any coherent explanation must include an explanation for why a given novel behavior came to be, making it available for selective forces to operate on.

The aim of this exercise has been twofold. First, it was to return the agency of human individuals and groups to nonabstract discussions of cultural evolution—organisms are not automatons blindly following programs, and human beings least of all. Humans actively and purposively engage with their physical and social environments to try to achieve what they perceive as potential net benefits, but they do so within the constraints of a given cultural system and what it deems acceptable to do. But, people will act in defiance of cultural constraints if they deem it necessary and if the benefits of defiance outweigh the social costs. If sufficient numbers do so, the system disintegrates and a new system that can accommodate the precipitating behaviors crystallizes around them.

Second, it was to lay out the structural framework of the proximate and ultimate processes that affect the expression of agential behaviors, in all its localized variability, and select from among them, using both the evolution of food production economies and the evolution of some of the typical components of complex societies as examples. However, that framework is first and foremost presented here as a heuristic device for modeling how the various aspects of culture dynamically interact to produce evolutionary culture change. Yes, real cultural systems are a good deal fuzzier both horizontally and vertically than the impression the heuristic device employed here conveys and the higher up the hierarchy of entities one looks, the fuzzier it gets. But then again, so is the species concept, even reproductively defined, and that does not prevent biologists from fruitfully employing it despite its fuzziness. For species, this is because in the final analysis there are things there whether or not our abstract definitions are sufficiently sophisticated to be able to isolate them precisely. The same applies to cultural and culturally structured systems. There are things there.

It is my hope that the model used here will prove useful for clarifying that there is more to the proverbial elephant than the part of its anatomy a given approach's magisterium addresses and how those various magisteria mesh. That is, it is my hope that this model will facilitate the explicit consideration of behavioral selection and how it can dovetail with the various understandings of the dynamics of cultural evolution and in so doing facilitate the synthesis of evolutionary culture theory that researchers have been working toward for almost as long as Darwinian approaches to the evolution of culture have been practiced.

References

Abu-Lughod, L. (1991). Writing against culture. In R. G. Fox (Ed.), *Recapturing anthropology: Working in the present* (pp. 137–162). SAR Press.

Adams, W. Y. (1998). *The philosophical roots of anthropology* (Center for the Study of language and information lecture notes no. 86). CSLI Publications.

Agelarakis, A. (1993). The Shanidar cave proto-Neolithic human population: Aspects of demography and paleopathology. *Human Evolution, 8*, 235–253.

Alt, K. W., Benz, M., et al. (2013). Earliest evidence for social endogamy in the 9000-year-old-population of Basta, Jordan. *PLoS One, 8*(6), e65649. https://journals.plos.org/plosone/article?id=10.1371/journal.pone.0065649

Anthony, D. W. (2007). *The horse, the wheel, and language: How bronze age riders from the Eurasian steppes shaped the modern world.* Princeton University Press.

Arnold, A. J., & Fristrup, K. (1982). The theory of evolution by natural selection: A hierarchical expansion. *Paleobiology, 8*, 113–129.

Arthur, W. B. (1990). Positive feedbacks in the economy. *Scientific American, 262*, 92–99.

Axelrod, R. (1984). *The evolution of cooperation*. Basic Books.

Bailey, R.C. (1991). *The behavioral ecology of Efe Pygmy men in the Ituri Forest, Zaire*. Museum of Anthropology Anthropological Papers No. 86. University of Michigan.

Barash, D. P. (1977). *Sociobiology and behavior*. Elsevior.

Barkow, J. H. (1989). Beyond the DP/DSS controversy. *Ethology and Sociobiology, 11*, 341–351.

Barkow, J. H., Cosmides, L., & Tooby, J. (Eds.). (1992). *The adapted mind: Evolutionary psychology and the generation of culture*. Oxford University Press.

Barrett, R. A. (1991). *Culture and conduct*. Wadsworth.

Bashkow, I. (2004). The neo-Boasian conception of cultural boundaries. *American Anthropologist, 106*, 443–458.

Bar-Yosef, O., & Belfer-Cohen, A. (1992). From foraging to farming in the Mediterranean Levant. In A. B. Gebauer & T. D. Price (Eds.), *Transitions to agriculture in prehistory (pp. 21–41). Monographs in world archaeology no. 4*. Prehistory Press.

Bar-Yosef, O., & Belfer-Cohen, A. (2010). Warfare in the levantine early Neolithic: A hypothesis to be considered. *Neo-Lithics, 1*(10), 6–10.

Belton, C. (2020). *Putin's people: How the KGB took back Russia and then took on the west*. Farrar, Straus and Giroux.

Berry, M. S. (1982). *Time, space, and transition in Anasazi prehistory*. University of Utah Press.

Bettinger, R. L. (1991). *Hunter-gatherers: Archaeological and evolutionary theory*. Plenum.

M. Rosenberg, *The Dynamics of Cultural Evolution*, Studies in Human Ecology and Adaptation 12, https://doi.org/10.1007/978-3-031-04863-0

Bettinger, R. L. (1998). Comment on M. Rosenberg, "Cheating at Musical Chairs: Sedentism in an Evolutionary Context". *Current Anthropology, 39*, 665–666.

Bettinger, R. L. (1999). From traveler to processor: Regional trajectories of hunter-gatherer sedentism in the Inyo-Mono region, California. In B. R. Billman & G. M. Feinman (Eds.), *Settlement pattern studies in the Americas: Fifty years since Viru* (pp. 39–55). Smithsonian Institution Press.

Bettinger, R. L. (2001). Holocene hunter-gatherers. In G. M. Feinman & T. D. Price (Eds.), *Archaeology at the millennium: A sourcebook* (pp. 137–195). Plenum.

Bettinger, R. L. (2009). Macroevolutionary theory and archaeology: Is there a big picture? In A. M. Prentiss, I. Kuijt, & J. C. Chatters (Eds.), *Macroevolution and human prehistory: Evolutionary theory and processual archaeology* (pp. 275–295). Springer.

Binford, L. R. (1968). Post Pleistocene adaptations. In S. R. Binford & L. R. Binford (Eds.), *New perspectives in archaeology* (pp. 313–341). Aldine.

Birch, J. (2017). *The philosophy of social evolution*. Oxford University Press.

Bird-David, N. (1992). Beyond "the original affluent society": A culturalist reformulation. *Current Anthropology, 33*, 25–47.

Blurton Jones, N. G. (1990). Three sensible paradigms for research on evolution and human behavior? *Ethology and Sociobiology, 11*, 353–359.

Bodet, C. (2019). Why are there no Neolithic mega-sites in the Anatolian Euphrates basin? A socio-archaeological approach to the marital structure of early farming societies. *The Journal of Anatolian Archaeological Studies, 2*, 21–51.

Boehm, C. (1993). Egalitarian behavior and reverse dominance hierarchy. *Current Anthropology, 34*, 227–254.

Boehm, C. (1999). *Hierarchy in the forest: The evolution of egalitarian behavior*. Harvard University Press.

Boehm, C. (2000). Conflict and the evolution of social control. *Journal of Consciousness Studies, 7*, 79–101.

Boggs, J. P. (2004). The culture concept as theory, in context. *Current Anthropology, 45*, 187–209.

Boone, J. L. (1992). Competition, conflict, and the development of social hierarchies. In E. A. Smith & B. Winterhalder (Eds.), *Evolutionary ecology and human behavior* (pp. 301–337). Aldine de Gruyter.

Boone, J. L., & Smith, E. A. (1998). Is it evolution yet: A critique of evolutionary archaeology. *Current Anthropology, 39*, S141–S173.

Borgerhoff-Mulder, M. (1990). Kipsigis women's preferences for wealthy men: Evidence for female choice in mammals. *Behavioral Ecology and Sociobiology, 27*, 255–264.

Boserup, E. (1965). *The conditions of agricultural growth*. Aldine.

Bowles, S., & Gintis, H. (2011). *A cooperative species: Human reciprocity and its evolution*. Princeton University Press.

Boyd, R., & Richerson, P. J. (1985). *Culture and the evolutionary process*. University of Chicago Press.

Boyd, R., & Richerson, P. J. (1987). The evolution of ethnic markers. *Cultural Anthropology, 2*, 65–79.

Boyd, R., & Richerson, P. J. (1990). Group selection among alternative evolutionarily stable strategies. *Journal of Theoretical Biology, 145*, 331–342.

Boyd, R., & Richerson, P. J. (1992). Punishment allows the evolution of cooperation (or anything else) in sizable groups. *Ethology and Sociobiology, 13*, 171–195.

Boyd, R., Mulder, M. B., Durham, W. H., & Richerson, P. J. (1997). Are cultural phylogenies possible? In P. Weingart et al. (Eds.), *Human by nature: Between biology and the social sciences* (pp. 355–386). Lawrence Erlbaum.

Brandon, R. N. (1988). Levels of selection: A hierarchy of interactors. In H. C. Plotkin (Ed.), *The role of behavior in evolution* (pp. 51–71). MIT Press.

Brandon, R. N. (1990). *Adaptation and environment*. Princeton University Press.

Brandt, E. (1994). Egalitarianism, hierarchy, and centralization in the pueblos. In W. Will & R. Leonard (Eds.), *The ancient southwestern community* (pp. 9–23). University of New Mexico.
Bromley, J., & Kozlov, V. (1989). The theory of ethnos and ethnic processes in soviet social sciences. *Comparative Studies of Society and History, 31*, 425–438.
Brumann, C. (1999). Writing for culture: Why a successful concept should not be discarded. *Current Anthropology, 40*(supplement), S1–S27.
Buss, D. M. (1994). *The evolution of desire*. Basic Books.
Byers, A. M. (2011). *Sacred games, death, and renewal in the ancient eastern woodlands: The Ohio Hopewell system of cult sodality heterarchies*. Altamira Press.
Cachel, S. (1989). The theory of punctuated equilibria and evolutionary anthropology. *Journal of Social and Biological Structures, 12*, 225–239.
Calloway, C. (1992). We have always been the frontier: The American revolution in Shawnee country. *American Indian Quarterly, 16*, 39–52.
Calloway, C. (2007). *The Shawnees and the war for America*. Penguin.
Carneiro, R. L. (1970). A theory of the origin of the state. *Science, 169*, 733–738.
Carneiro, R. L. (1973). Structure, function, and equilibrium in the evolutionism of Herbert Spencer. *Journal of Anthropological Research, 29*, 77–95.
Carneiro, R. L. (1998). What happened at the flashpoint: Conjectures on chiefdom formation at the very moment of conception. In E. M. Redmond (Ed.), *Chiefdoms and chieftaincy in the Americas* (pp. 18–42). University Press of Florida.
Cavalli-Sforza, L. L., & Feldman, M. W. (1981). *Cultural transmission and evolution: A quantitative approach*. Princeton University Press.
Chagon, N. A., & Bugos, P. E., Jr. (1979). Kin selection and conflict: An analysis of a Yanomamö axe fight. In N. A. Chagnon & W. Irons (Eds.), *Evolutionary biology and human social behavior: An anthropological perspective* (pp. 213–238). North Scituate.
Charles, D. K., & Buikstra, J. E. (1983). Archaic mortuary sites in the Central Mississippi drainage: Distribution, structure and behavioral implications. In J. L. Phillips & J. A. Brown (Eds.), *Archaic hunters and gatherers in the American Midwest* (pp. 117–145). Academic Press.
Chase, A. K. (1989). Domestication and domiculture in northern Australia: A social perspective. In D. R. Harris & G. C. Hillman (Eds.), *Foraging and farming: The evolution of plant exploitation* (pp. 42–54). Unwin Hyman.
Chatters, J. C. (2009). A macroevolutionary perspective on the archaeological record of North America. In A. M. Prentiss, I. Kuijt, & J. C. Chatters (Eds.), *Macroevolution and human prehistory: Evolutionary theory and processual archaeology* (pp. 213–234). Springer.
Chatters, J. C., & Prentiss, W. C. (2005). A Darwinian macroevolutionary perspective on the development of hunter-gatherer systems in northwestern North America. *World Archaeology, 37*, 46–65.
Childe, V. G. (1936). *Man makes himself*. Watts and Co.
Clark, J. E. (1993). *The Shawnee*. University Press of Kentucky.
Clarke, D. L. (1968). *Analytical archaeology*. Methuen.
Cohen, M. N. (1977). *The food crisis in prehistory: Overpopulation and the origins of agriculture*. Yale University Press.
Cooley, C. H. (1909). *Social organization: A study of the larger mind*. Charles Scribner's Sons.
Cosmides, L., Tooby, J., & Barkow, J. H. (1992). Introduction: Evolutionary psychology and conceptual integration. In J. H. Barkow et al. (Eds.), *The adapted mind: Evolutionary psychology and the generation of culture* (pp. 3–15). Oxford University Press.
Cronk, L. (1999). *That complex whole: Culture and the evolution of human behavior*. Westview Press.
Cronk, L. (2006). Intelligent design in cultural evolution. Comment on Masoudi et al, "towards a unified science of cultural evolution". *Behavioral and Brain Sciences, 29*, 352–353.
Crowley, R. (2011). *City of fortune: How Venice ruled the seas*. Random House.
Damuth, J. (1985). Selection among species: A formulation in terms of natural functional units. *Evolution, 39*, 1132–1146.

Damuth, J., & Heisler, J. L. (1988). Alternative formulations of multilevel selection. *Biology and Philosophy, 3*, 407–430.
Darwin, C. (1871). *The descent of man and evolution in relation to sex*. Murray.
Dawkins, R. (1976). *The selfish gene*. Oxford University Press.
Dawkins, R. (1982). *The extended phenotype: The long reach of the gene*. Oxford University Press.
Dawkins, R. (1986). *The blind watchmaker*. Longham.
Dawkins, R. (1989). *The selfish gene (new edition)*. Oxford University Press.
Dayan, D. I., Graham, M. A., Baker, J. A., & Foster, S. A. (2019). Incorporating the environmentally sensitive phenotype into evolutionary thinking. In T. Uller & K. N. Laland (Eds.), *Evolutionary causation: Biological and philosophical reflections* (pp. 81–107). MIT Press.
Degler, C. N. (1991). *In search of human nature: The decline and revival of Darwinism in American social thought*. Oxford University Press.
De Vries, H. (1905). *Species and varieties: Their origin by mutation*. Open Court Publishing.
De Vries, H. (1909). *The mutation theory. D.* Appleton and Co.
De Vries, H. (1910). *Intracellular pangenesis*. Open Court Publishing Co.
De Waal, F. (1982). *Chimpanzee politics: Power and sex among apes*. Jonathan Cape.
De Waal, F. (1996). *Good natured: The origins of right and wrong in humans and other animals*. Harvard University Press.
Diamond, J. (1997). *Guns, germs, and steel: The fates of human societies*. W.W. Norton.
Diener, P. (1980). Quantum adjustment, macroevolution, and the social field: Some comments on evolution and culture. *Current Anthropology, 21*(4), 423–431.
Dobzhansky, T. (1951). *Genetics and the origin of species (third edition)*. Columbia University Press.
DuBois, C. (1955). The dominant value profile of American culture. *American Anthropologist, 57*, 1232–1239.
Dugatkin, L. A., & Reeve, H. K. (1994). Behavioral ecology and levels of selection. *Advances in the Study of Behavior, 23*, 101–133.
Dumond, D. E. (1972). Population growth and political centralization. In B. Spooner (Ed.), *Population growth: Anthropological implications* (pp. 286–310). MIT Press.
Dunnell, R. C. (1978). Style and function: A fundamental dichotomy. *American Antiquity, 43*, 192–202.
Dunnell, R. C. (1980). Evolutionary theory and archaeology. In M. B. Schiffer (Ed.), *Advances in archaeological method and theory* (Vol. 3, pp. 38–99). Academic Press.
Dunnell, R. C. (1986). Methodological issues in americanist artifact classification. In M. B. Schiffer (Ed.), *Advances in archaeological method and theory* (Vol. 9, pp. 149–207). Academic Press.
Dunnell, R. C. (1989). Aspects of the application of evolutionary theory in archaeology. In C. C. Lamberg-Karlovski (Ed.), *Archaeological thought in America* (pp. 35–49). Cambridge University Press.
Dunnell, R. C. (1995). What is it that actually evolves? In P. A. Teltser (Ed.), *Evolutionary archaeology: Methodological issues* (pp. 33–50). University of Arizona Press.
Durham, W. H. (1979). Toward a coevolutionary theory of human biology and culture. In N. A. Chagnon & W. Irons (Eds.), *Evolutionary biology and human social behavior: An anthropological perspective* (pp. 39–59). Duxbury Press.
Durham, W. H. (1990). Advances in evolutionary culture theory. *Annual Review of Anthropology, 19*, 187–210.
Durham, W. H. (1991). *Coevolution: Genes, culture, and human diversity*. Stanford University Press.
Dyson-Hudson, R., & Smith, E. A. (1978). *Human territoriality: An ecological reassessment: American Anthropologist, 80*, 21–41.
Eldredge, N. (1971). The allopatric model and phylogeny in Paleozoic invertebrates. *Evolution, 18*, 156–167.
Eldredge, N. (1985a). *Unfinished synthesis: Biological hierarchies and modern evolutionary thought*. Oxford University Press.

Eldredge, N. (1985b). *Time frames: The rethinking of darwinian evolution and the theory of punctuated equilibria*. Simon and Schuster.
Eldredge, N. (1989). *Macroevolutionary patterns and evolutionary dynamics: Species, niches and adaptive peaks*. McGraw-Hill.
Eldredge, N. (1999). *The pattern of evolution*. Freeman.
Eldredge, N. (2002). Hierarchy: Theory and praxis in evolutionary biology. In R. E. Auxier & L. E. Hahn (Eds.), *The philosophy of Marjorie Grene. With reply by Marjorie Grene* (The Library of Living Philosophers XXIX) (pp. 315–334). Open Court Publishing Co.
Eldredge, N. (2003). Human triangles: Genes, sex, and economics in human evolution. In S. J. Scher & F. Rauscher (Eds.), *Evolutionary psychology: Alternative approaches* (pp. 91–110). Springer.
Hahn, L.E (eds), *The philosophy of Marjorie Grene. With reply by Majorie Grene* (315–334). The Library of Living Philosophers XXIX. Open Court Publishing Co.
Eldredge, N. (2015). *Eternal ephemera: Adaptation and the origin of species from the nineteenth century through punctuated equilibria and beyond*. Columbia University Press.
Eldredge, N., & Cracraft, J. (1980). *Phylogenetic patterns and the evolutionary process: Method and theory in comparative biology*. Columbia University Press.
Eldredge, N., & Gould, S. J. (1972). Punctuated equilibria: An alternative to phyletic gradualism. In T. J. M. Schopf (Ed.), *Models in paleobiology* (pp. 82–115). Freeman, Cooper.
Eldredge, N., & Salthe, S. N. (1984). Hierarchy and evolution. *Oxford Surveys in Evolutionary Biology, 1*, 182–206.
Etkin, W. (1967). *Social behavior from fish to man*. University of Chicago Press.
Field, J. S. (2004). Environmental and climatic considerations: A hypothesis for conflict and the emergence of social complexity in Fijian prehistory. *Journal of Anthropological Archaeology, 23*, 79–99.
Fisher, R. A. (1958). *The genetical theory of evolution (second edition)*. Dover.
Flannery, K. V. (1972). The origin of the village as a settlement type in Mesoamerica and the Near East. In P. J. Ucko, R. Tringham, & G. W. Dimbleby (Eds.), *Man, settlement and urbanism* (pp. 23–53). Duckworth.
Flanagan, J. G. (1989). Hierarchy in simple "egalitarian" societies. *Annual Review of Anthropology, 18*, 245–266.
Foster, G. (1960). *Culture and conquest: America's Spanish heritage* (Viking fund publications in anthropology, #27). Wenner-Gren.
Geertz, C. (1973). *The interpretation of cultures*. Basic Books.
Gibson, J. J. (1979). *The ecological approach to visual perception*. Houghton Mifflin.
Ghiselin, M. T. (1974). A radical solution to the species problem. *Systematic Zoology, 23*, 536–544.
Gladwin, T. (1957). Personality structure in the plains. *Anthropological Quarterly., 30*, 111–124.
Godfrey-Smith, P. (2009). *Darwinian populations and natural selection*. Oxford University Press.
Goldschmidt, R. (1940). *The material basis of evolution*. Yale University Press.
Goodenough, O. R., & Dawkins, R. (1994). The St. Jude mind virus. *Nature, 371*, 23–24.
Goodenough, W. H. (1981). *Culture, language, and society* (2nd ed.). Benjamin/Cummings.
Goodenough, W. H. (1999). Outline of a framework for a theory of cultural evolution. *Cross-Cultural Research, 35*, 84–107.
Gould, S. J. (1976). Ladders, bushes, and human evolution. *Natural History, 85*(4), 24–31.
Gould, S. J. (1977). *Ontogeny and phylogeny*. Harvard University Press.
Gould, S. J. (1980). Is a new and general theory of evolution emerging? *Paleobiology, 6*, 119–130.
Gould, S. J. (1982a). The meaning of punctuated equilibrium and its role in validating a hierarchical approach to macroevolution. In R. Milkman (Ed.), *Perspectives on evolution* (pp. 83–104). Sinauer.
Gould, S. J. (1982b). Darwinism and the expansion of evolutionary theory. *Science, 216*, 380–387.
Gould, S. J. (1983). *Hen's teeth and horse's toes: Further reflections on natural history*. W.W. Norton.
Gould, S. J. (1987). The panda's thumb of technology. *Natural History, 96*, 14–23.

Gould, S. J. (1989a). Punctuated equilibria in fact and theory. *Journal of Social and Biological Structures, 12*, 241–250.

Gould, S. J. (1989b). *Wonderful life: The burgess shale and the nature of history*. W.W. Norton.

Gould, S. J. (2002a). *The structure of evolutionary theory*. Belknap.

Gould, S. J. (2002b). *I have landed: The end of a beginning in natural history*. Harmony Books.

Gould, S. J., & Eldredge, N. (1977). Punctuated equilibria: The tempo and mode of evolution reconsidered. *Paleobiology, 3*, 115–151.

Gould, S. J., & Lewontin, R. C. (1979). The spandrels of San Marco and the panglossian paradigm: A critique of the adaptationist programme. *Proc. Royal Society London Series B, 205*, 581–598.

Grantham, T. A. (1995). Hierarchical approaches to macroevolution: Recent work on species selection and the "Effect Hypothesis". *Annual Review of Ecology and Systematics, 26*, 301–321.

Grossman, L. (2010). Prehistoric warfare – Cause and visibility. *Neo-Lithics, 1*(10), 36–37.

Hall, B. K. (1992). *Evolutionary developmental biology*. Chapman and Hall.

Hall, B. K. (2012). Evolutionary developmental biology (Evo-devo): Past, present, and future. *Evolution: Education and Outreach, 5*, 184–193.

Hamilton, W. D. (1971). Selection of selfish and altruistic behavior in some extreme models. In J. E. Eisenberg & W. S. Dillon (Eds.), *Man and beast: Comparative social behavior* (pp. 57–91). Smithsonian Institution Press.

Hamilton, W. D. (1987). Discriminating nepotism: Expectable, common, overlooked. In D. S. C. Fletcher & C. D. Michener (Eds.), *Kin recognition in animals* (pp. 417–437). John Wiley.

Handwerker, W. P. (2002). The construct validity of cultures: Cultural diversity, culture theory, and a method for ethnography. *American Anthropologist, 104*, 106–122.

Hanson, F. A. (2004). The new superorganic. *Current Anthropology, 45*, 467–482.

Harris, M. (1968). *The rise of anthropological theory*. Thomas Y. Crowell Co.

Harris, M. (1979). *Cultural materialism: The struggle for a science of culture*. Random House.

Harris, M. (1994). Cultural materialism is alive and well and won't go away until something better comes along. In R. Borofsky (Ed.), *Assessing cultural anthropology* (pp. 62–75). McGraw-Hill.

Harris, M. (1999). *Theories of culture in postmodern times*. AltaMira.

Hendry, A. P., Wenberg, J. K., Bentzen, P., Volk, E. C., & Quinn, T. P. (2000). Rapid evolution of reproductive isolation in the wild: Evidence from introduced salmon. *Science, 290*, 516–518.

Henrich, J. (2004). Cultural group selection, Coevolutionary processes and large-scale cooperation. *Journal of Economic Behavior and Organization, 53*, 3–35.

Henry, D. O. (1989). *From foraging to agriculture: The Levant at the end of the ice age*. University of Pennsylvania Press.

Hewlett, B. S., & Cavalli-Sforza, L. L. (1986). Cultural transmission among Aka pygmies. *American Anthropologist, 88*, 922–934.

Hoffman, A. (1989). Fifteen years later: Punctuated equilibria in retrospect. *Journal of Social and Biological Structures, 12*, 185–194.

Hsu, F. L. K. (1961). American core values and national character. In F. L. K. Hsu (Ed.), *Psychological anthropology* (pp. 209–230). Dorsey press.

Hull, D. L. (1976). Are species really individuals? *Systematic Zoology, 25*, 174–191.

Hull, D. L. (1978). A matter of individuality. *Philosophy of Science, 45*, 335–360.

Hull, D. L. (1980). Individuality and selection. *Annual Review of Ecology and Systematics, 11*, 311–332.

Hull, D. L. (1988). *Science as a process: An evolutionary account of the social and conceptual development of science*. University of Chicago Press.

Hull, D. L. (1994). Taking vehicles seriously. *Behavioral and Brain Sciences, 17*, 627–628.

Hull, D. L. (2001). *Science and selection: Essays on biological evolution and the philosophy of science*. Cambridge University Press.

Hull, D. L., Langman, R. E., & Glenn, S. S. (2001). A general account of selection: Biology, immunology and behavior. *Behavioral and Brain Sciences, 24*, 511–573.

Irons, W., & Cronk, L. (2000). Two decades of a new paradigm. In L. Cronk, N. Chagnon, & W. Irons (Eds.), *Adaptation and human behavior* (pp. 3–26). Aldine de Gruyter.

Johnson, A. W., & Earle, T. (2000). *The evolution of human societies: From foraging group to agrarian state (second edition)*. Stanford University Press.

Jones, G. T., Leonard, R. D., & Abbott, A. L. (1995). The structure of selectionist explanations in archaeology. In P. A. Teltser (Ed.), *Evolutionary archaeology: Methodological issues* (pp. 13–32). University of Arizona Press.

Joyce, R. (2006). *The evolution of morality*. MIT Press.

Keeley, L. H. (1996). *War before civilization: The myth of the peaceful savage*. Oxford University Press.

Kelly, R. C. (2005). The evolution of lethal intergroup violence. *Proc. Natl. Acad. Sci. USA, 102*(43), 15294–15298.

Kelly, R. L. (1983). Hunter-gatherer mobility strategies. *Journal of Anthropological Research, 39*, 277–306.

Kemp, B. M., Judd, K., Munroe, C., Eerkens, J. W., et al. (2017). Prehistoric mitochondrial DNA of domesticate animals supports a 13th century exodus from the northern US southwest. *PLoS One, 12*(7), e0178882. https://journals.plos.org/plosone/article?id=10.1371/journal.pone.0178882

Kirch, P. V. (1984). *The evolution of polynesian chiefdoms*. Cambridge University Press.

Kohler, T. A., & Gumerman, G. J. (Eds.). (2000). *Dynamics in human and primate societies: Agent based modeling of social and spatial processes*. Oxford University Press.

Krebs, J. R. (1978). Optimal foraging: Decision rules for predators. In J. R. Krebs & N. B. Davies (Eds.), *Behavioral ecology: An evolutionary approach* (pp. 23–63). Blackwell.

Kroeber, A. (1917). The superorganic. *American Anthropologist, 19*, 163–213.

Kroeber, A. L., & Kluckhohn, C. (1952). *Culture: A critical review of concepts and definitions*. Vintage Books.

Kuhn, T. S. (1962). *The structure of scientific revolutions*. University of Chicago Press.

Kuijt, I., & Prentiss, A. M. (2009). Niche construction, macroevolution, and the late Epipaleolithic of the near east. In A. M. Prentiss, I. Kuijt, & J. C. Chatters (Eds.), *Macroevolution and human prehistory: Evolutionary theory and processual archaeology* (pp. 253–271). Springer.

Laland, K. N., & Brown, G. R. (2002). *Sense and nonsense: Evolutionary perspectives on human behavior*. Oxford University Press.

Laland, K. N., Odling-Smee, J., & Feldman, M. C. (2019). Understanding niche construction as an evolutionary process. In T. Uller & K. N. Laland (Eds.), *Evolutionary causation: Biological and philosophical reflections* (pp. 127–152). MIT Press.

Layton, R., Foley, R., & Williams, E. (1991). The transition between hunting and gathering and the specialized husbandry of resources. *Current Anthropology, 32*, 255–274.

LeBlanc, S. A. (2003). *Constant battles*. St. Martin's Press.

LeBlanc, S. A. (2010). Early Neolithic warfare in the near east and its broader implications. *Neo-Lithics, 1*(10), 40–49.

Lee, R.B. (1965). *The subsistence ecology of the !Kung bushmen*. Ph.D. dissertation, University of California, Berkeley.

Lee, R. B. (1968). What hunters do for a living, or, how to make out on scarce resources. In R. B. Lee & I. DeVore (Eds.), *Man the hunter* (pp. 30–48). Aldine.

Lee, R. B. (1972). The intensification of social life among the !Kung bushmen. In B. Spooner (Ed.), *Population growth: Anthropological implications (pp. 343–350)*. MIT Press.

Lee, R. B. (1979). *The !Kung san: Men, women, and work in a foraging society*. Cambridge University Press.

Lee, R. B., & DeVore, I. (Eds.). (1968). *Man the hunter*. Aldine.

Leonard, R. D., & Jones, G. T. (1987). Elements of an inclusive evolutionary model for archaeology. *Journal of Anthropological Archaeology, 6*, 199–219.

Levine, R. (1967). The internalization of political values in stateless societies. In R. Hunt (Ed.), *Personalities and cultures: Readings in psychological anthropology* (pp. 185–203). Natural History Press.
Lévi-Strauss, C. (1982). *The way of the masks (trans: Modelski, S.)*. University of Washington Press.
Lewontin, R. C. (1970). The units of selection. *Annual Review of Ecology and Systematics, 1*, 1–18.
Lewontin, R. C. (1982). Organism and environment. In H. C. Plotkin (Ed.), *Learning, development and culture: Essays in evolutionary epistemology* (pp. 151–170). Wiley.
Lewontin, R. C. (1983). Gene, organisms, and environment. In D. S. Bendall (Ed.), *Evolution from molecules to men* (pp. 273–285). Cambridge University Press.
Lewontin, R. C. (2001). Genes, organisms and environments. In S. Oyama et al. (Eds.), *Cycles of contingency: Developmental systems and evolution* (pp. 59–66). MIT Press.
Lloyd, E. A. (1988). *The structure and confirmation of evolutionary theory*. Greenwood Press.
Lowie, R. (1973). Political organization among American Indians. In D. Black & M. Mileski (Eds.), *Social organization of law* (pp. 278–303). Seminar Press.
Lyman, R. L., & O'Brien, M. J. (1998). The goals of evolutionary archaeology: History and explanation. *Current Anthropology, 39*, 615–652.
MacMahorn, J. A., Phillips, D. L., Robinson, J. V., & Schimpf, D. J. (1978). Levels of biological organization: An organism centered approach. *Bioscience, 28*, 700–704.
Mace, R., Holden, C. J., & Shennan, S. (Eds.). (2005). *The evolution of cultural diversity: A phylogenetic approach*. Left Coast Press.
Maisels, C. K. (1999). *Early civilizations of the Old World: The formative histories of Egypt, the Levant, Mesopotamia, India and China*. Rouledge.
Malthus, T.R. (1798). *An essay on the principle of population as it affects the future improvement of society with remarks on the speculations of Mr. Godwin, M. Condorcet, and other writers*. J. Johnson.
Maxwell, T. D. (1995). The use of comparative and engineering analyses in the study of prehistoric agriculture. In P. A. Teltser (Ed.), *Evolutionary archaeology: Methodological issues* (pp. 113–128). University of Arizona Press.
Maybury-Lewis, D. (1974). *Akwe-Shavante society*. Oxford University Press.
Maynard Smith, J. (1982). *Evolution and the theory of games*. Cambridge University Press.
Maynard Smith, J. (1989). J. Weismann and modern biology. In P. H. Harvey & L. Partridge (Eds.), *Oxford surveys in evolutionary biology* (pp. 1–12). Oxford University Press.
Mayr, E. (1942). *Systematics and the origin of species*. Columbia University Press.
Mayr, E. (1954). Change of genetic environment and evolution. In J. Huxley, A. C. Hardy, & E. B. Ford (Eds.), *Evolution as a process* (pp. 157–180). Allen and Unwin.
Mayr, E. (1961). Cause and effect in biology. *Science, 134*, 1501–1506.
Mayr, E. (1963). *Animal species and evolution*. Harvard University Press.
Mayr, E. (1982). Speciation and macroevolution. *Evolution, 36*, 1119–1132.
Mayr, E. (1989). Speciational evolution or punctuated equilibria. *Journal of Social and Biological Structures, 12*, 137–158.
Mesoudi, A., Whiten, A., & Laland, K. N. (2006). Towards a unified science of cultural evolution. *Behavioral and Brain Sciences, 29*, 329–383.
Miller, G. F. (1999). Sexual selection for cultural displays. In R. Dunbar, C. Knight, & C. Power (Eds.), *The evolution of culture* (pp. 71–91). Rutgers University Press.
Miller, R. J. (1964). Cultures as religious structures. In J. Helm (Ed.), *Symposium on new approaches to the study of religion: Proceedings of the 1964 annual spring meeting of the American ethnological society*. University of Washington Press.
Morgan, L. H. (1851). *League of the Ho-De-No-Sau-Nee, or Iroquois*. Rochester.
Nakao, H., Tamura, K., Arimatsu, Y., Nakagawa, T., Matsumoto, N., & Matsugi, T. (2016). Violence in the prehistoric period of Japan: The spatio-temporal pattern of skeletal evidence for violence in the Jomon period. *Biology Letters, 12*, 20160028. https://doi.org/10.1098/rsbl.2016.0028

O'Brien, M. J., & Lyman, R. L. (2000). *Applying evolutionary archaeology: A systemic approach*. Kluwer/Plenum.

O'Brien, M. J., Lyman, R. L., & Leonard, R. D. (2003). What is evolution: A response to Bamforth. *American Antiquity, 68*, 573–580.

Odling-Smee, F. J. (1988). Niche constructing phenotypes. In H. C. Plotkin (Ed.), *The role of behavior in evolution* (pp. 73–132). Cambridge.

Odling-Smee, F. J., Laland, K. N., & Feldman, M. W. (2003). *Niche construction: The neglected process in evolution*. Princeton University Press.

Oliver, S. C. (1962). *Ecology and cultural continuity as contributing factors in the social organization of the plains Indians* (Vol. 48, p. 1). University of California Publications in American Archaeology and Ethnology.

Okasha, S. (2018). *Agents and goals in evolution*. Oxford University Press.

Özkaya, V., & Coşkun, A. (2011). Körtik Tepe. In M. Özdoğan, N. Başgelen, & P. Kuniholm (Eds.), *The Neolithic in Turkey: New excavations & new research, the Tigris Basin* (pp. 89–127). Archaeology and Art Publications.

Pagel, M. (2006). Darwinian cultural evolution rivals genetic evolution. Comment on Masoudi et al, "towards a unified science of cultural evolution.". *Behavioral and Brain Sciences, 29*, 360.

Paterson, H. E. H. (1981). The continuing search for the unknown and unknowable: A critique of contemporary ideas on speciation. *South African Journal of Science, 77*, 113–119.

Paterson, H. E. H. (1985). The recognition concept of species. In E. S. Vrba (Ed.), *Species and speciation (pp. 21–29)* (Vol. 4). Transvaal Mus. Monographs.

Pinker, S. (2002). *The blank slate: The modern denial of human nature*. Penguin.

Prentiss, A. M. (2009). The emergence of new socioeconomic strategies in the middle and late Holocene Pacific northwest region of North America. In A. M. Prentiss, I. Kuijt, & J. C. Chatters (Eds.), *Macroevolution and human prehistory: Evolutionary theory and processual archaeology* (pp. 111–131). Springer.

Prentiss, W. C., & Chatters, J. C. (2003). Cultural diversification and decimation in the prehistoric record. *Current Anthropology, 44*, 33–58.

Prentiss, A. M., Kuijt, I., & Chatters, J. C. (Eds.). (2009). *Macroevolution and human prehistory: Evolutionary theory and processual archaeology*. Springer.

Price, T. D. (1983). The European Mesolithic. *American Antiquity, 48*, 761–778.

Price, T. D. (1985). Affluent foragers of Mesolithic southern Scandinavia. In T. D. Price & J. A. Brown (Eds.), *Prehistoric hunter-gatherers: The emergence of cultural complexity* (pp. 341–363). Academic Press.

Radcliffe-Brown, A. R. (1952). *Structure and function in primitive society*. Oxford University Press.

Radcliffe-Brown, A. R. (1957). *A natural science of society*. University of Chicago Press.

Rafferty, J. E. (1985). The archaeological record on sedentariness: Recognition, development, and implications. In M. B. Schiffer (Ed.), *Advances in archaeological method and theory* (Vol. 8, pp. 113–156). Academic Press.

Rappaport, R. A. (1968). *Pigs for the ancestors: Ritual in the ecology of a New Guinea people*. Yale University Press.

Renfrew, C. (1978). The anatomy of innovation. In D. Green, C. Haselgrove, & M. Spriggs (Eds.), *Social organisation and settlement: Contributions from anthropology, archaeology and geography* (pp. 89–117) BAR International Series 47(1).

Richerson, P. J., & Boyd, R. (1999). Complex societies: The evolutionary origins of a crude superorganism. *Human Nature, 10*, 253–289.

Richerson, P. J., Boyd, R., & Bettinger, R. L. (2001). Was agriculture impossible during the Pleistocene but mandatory during the Holocene? A climate change hypothesis. *American Antiquity, 66*, 387–411.

Rindos, D. (1984). *The origin of agricultural systems: An evolutionary perspective*. Academic Press.

Rindos, D. (1985). Darwinian selection, symbolic variation, and the evolution of culture. *Current Anthropology, 26*, 65–77.

Roscoe, P. (2002). Culture. In J. P. Hart & J. E. Terrell (Eds.), *Darwin and archaeology: A handbook of key concepts* (pp. 107–124). Bergin and Garvey.

Rosenberg, M. (1990). The mother of invention: Evolutionary theory, territoriality, and the origins of agriculture. *American Anthropologist, 92*, 399–415.

Rosenberg, M. (1994). Pattern, process, and hierarchy in the evolution of culture. *Journal of Anthropological Archaeology, 13*, 307–340.

Rosenberg, M. (1998). Cheating at musical chairs: Sedentism in an evolutionary context. *Current Anthropology, 39*, 653–681.

Rosenberg, M. (2003). The strength of numbers: From villages to towns in the Aceramic Neolithic of southwestern Asia. In M. Özdoğan et al. (Eds.), *From primary villages to cities: Studies presented to Ufuk Esin* (pp. 91–101). Arkeoloji ve Sanat Yayınları.

Rosenberg, M. (2009). Proximate causation, group selection, and the evolution of hierarchical human societies: System, process, and pattern. In A. M. Prentiss, I. Kuijt, & J. C. Chatters (Eds.), *Macroevolution and human prehistory: Evolutionary theory and processual archaeology* (pp. 23–49). Springer.

Rosenberg, M. (2011). Hallan Çemi. In M. Özdoğan, N. Başgelen, & P. Kuniholm (Eds.), *The Neolithic in Turkey: New excavations & new research, the Tigris Basin* (pp. 61–78). Archaeology and Art Publications.

Rosenberg, M., & Rocek, T. R. (2019). Socio-political organization in the Aceramic Neolithic of southwestern Asia: The complex evolution of socio-political complexity. *Journal of Anthropological Archaeology, 54*, 17–30.

Ruse, M. (1989). Is the theory of punctuated equilibria a new paradigm? *Journal of Social and Biological Structures, 12*, 195–212.

Sahlins, M. D. (1960). Evolution: Specific and general. In M. D. Sahlins & E. R. Service (Eds.), *Evolution and culture* (pp. 12–44). University of Michigan Press.

Schmidt, K. (2012). *Göbekli Tepe: A stone age sanctuary in South-Eastern Anatolia*. Ex oriente.

Schuman, R. (1950). Speech of May 9, 1950 ('the Schuman declaration').

Service, E. R. (1960). The law of evolutionary potential. In M. D. Sahlins & E. R. Service (Eds.), *Evolution and culture* (pp. 93–122). University of Michigan Press.

Service, E. R. (1962). *Primitive social organization*. Random House.

Shennan, S. (2000). Population, culture history, and the dynamics of culture change. *Current Anthropology, 41*, 811–835.

Shennan, S. (2002). *Genes, memes and human history*. Thames and Hudson.

Shennan, S. J., & Steele, J. (1999). Cultural learning in hominids: A behavioral ecological approach. In H. Box & K. Gibson (Eds.), *Mammalian social learning* (pp. 367–388). Cambridge University Press.

Shostak, S. (2003). Struggle in the structure of evolutionary theory. *Boundary 2, 30*(3), 213–239.

Simpson, G. G. (1944). *Tempo and mode in evolution*. Columbia University Press.

Simpson, G. G. (1953). *The major features of evolution*. Columbia University Press.

Skinner, B. F. (1981). Selection by consequences. *Science, 213*, 501–504.

Slater, P. (1970). *The pursuit of loneliness*. Beacon Press.

Smith, P. (1973). Family burials at Hayonim: A brief communication. *Paléorient, 1*, 69–71.

Smith, A. D. (1986). *The ethnic origin of nations*. Blackwell.

Smith, E. A. (1988). Risk and uncertainty in the (original affluent society): Evolutionary ecology of resource-sharing and land tenure. In T. Ingold, D. Riches, & J. Woodburn (Eds.), *Hunters and gatherers: History, evolution and social change* (pp. 222–252). Berg.

Smith, A. D. (2000a). *The nation in history: Historiographic debates about ethnicity and nationalism*. University Press of New England.

Smith, E. A. (2000b). Three styles in the evolutionary analysis of human behavior. In L. Cronk et al. (Eds.), *Adaptation and human behavior* (pp. 27–46). Aldine de Gruyter.

Sober, E. (1984). *The nature of selection: Evolutionary theory in philosophical focus*. University of Chicago Press.

Sober, E., & Wilson, D. S. (1998). *Unto others: The evolution and psychology of unselfish behavior*. Harvard University Press.

Soltis, J., Boyd, R., & Richerson, P. J. (1995). Can group-functional behaviors evolve by cultural group selection: An empirical test. *Current Anthropology, 36*, 473–494.

Spencer, C. S. (1987). Rethinking the chiefdom. In R. D. Drennan & C. Uribe (Eds.), *Chiefdoms in the Americas* (pp. 1–9). Cambridge University Press.

Spencer, C. S. (1990). On the tempo and mode of state formation: Neoevolutionism reconsidered. *Journal of Anthropological Archaeology, 9*, 1–30.

Spencer, C. S. (1993). Human agency, biased transmission, and the cultural evolution of chiefly authority. *Journal of Anthropological Archaeology, 12*, 41–74.

Spencer, C. S. (1997). Evolutionary approaches in archaeology. *Journal of Archaeological Research, 5*, 209–264.

Spencer, C. S. (1998). A mathematical model of primary state formation. *Cultural Dynamics, 10*, 5–20.

Spencer, C. S. (2009). Testing the morphogenesist model of primary state formation: The Zapotec case. In A. M. Prentiss, I. Kuijt, & J. C. Chatters (Eds.), *Macroevolution and human prehistory: Evolutionary theory and processual archaeology* (pp. 133–155). Springer.

Spencer, H. (1857). Progress: Its laws and causes. *Westminster Review, 67*, 445–485.

Spikins, P. (2008). The bashful and the boastful: Prestigious leaders and social change in mesolithic societies. *Journal of World Prehistory, 21*(4), 173–193.

Stanish, C., & Haley, K. J. (2005). Power, fairness, and architecture: Modeling early chiefdom development in the Central Andes. In K. J. Vaughn et al. (Eds.), *Foundations of power in the prehispanic Andes* (pp. 53–70). Archeological Papers of the American Anthropological Association 14.

Stanley, S. M. (1975). A theory of evolution above the species level. *Proceedings of the National Academy of Science, 72*, 646–650.

Stanley, S. M. (1979). *Macroevolution: Pattern and process*. W.H. Freeman.

Stanley, S. M. (1981). *The new evolutionary timetable: Fossils, genes, and the origin of species*. Basic Books.

Stanley, S. M. (1989). The empirical case for the punctuational model of evolution. *Journal of Social and Biological Structures, 12*, 159–172.

Sterelny, K. (2001). *Dawkins vs Gould: Survival of the fittest*. Totem Books.

Steward, J. H. (1938). *Basin plateau aboriginal sociopolitical groups* (Vol. 120). Bureau of American Ethnology Bulletin.

Steward, J. H. (1955). *Theory of culture change: The methodology of multilinear evolution*. University of Illinois Press.

Symons, D. (1989). A critique of Darwinian anthropology. *Ethology and Sociobiology, 10*, 131–144.

Symons, D. (1992). On the use and misuse of Darwinism in the study of human behavior. In J. H. Barkow et al. (Eds.), *The adapted mind: Evolutionary psychology and the generation of culture* (pp. 137–159). Oxford University Press.

Tainter, J. A. (1988). *The collapse of complex societies*. Cambridge University Press.

Taylor, W. W. (1948). *A study of archaeology*. American Anthropological Association. *Memoir, 69*.

Tuchman, B. A. (1978). *A distant mirror*. Alfred A, Knopf.

Tooby, J., & Cosmides, L. (1990). The past explains the present: Emotional adaptations and the structure of ancestral environments. *Ethology and Sociobiology, 11*, 375–424.

Tooby, J., & Cosmides, L. (1992). The psychological foundations of culture. In J. H. Barkow et al. (Eds.), *The adapted mind: Evolutionary psychology and the generation of culture* (pp. 19–136). Oxford University Press.

Turke, P. W. (1990). Which humans behave adaptively, and why does it matter. *Ethology and Sociobiology, 11*, 305–339.

Tylor, E. B. (1871). *Primitive culture*. Murray.

Uller, T., & Laland, K. N. (2019). Evolutionary causation. In T. Uller & K. N. Laland (Eds.), *Evolutionary causation: Biological and philosophical reflections* (pp. 1–12). MIT Press.

Valentine, J. W. (1973). *Evolutionary paleoecology of the marine biosphere*. Prentiss-Hall.

Van Sommers, P. (1972). *The biology of behavior*. John Wiley & Sons.

Varisco, D. M. (2018). *Culture still matters: Notes from the field*. Brill.

Veblen, T. (1915). *Imperial Germany and the industrial revolution*. University of Michigan Press.

Vrba, E. S. (1980). Evolution, species and fossils: How does life evolve. *South African Journal of Science, 76*, 61–84.

Vrba, E. S. (1983). Macroevolutionary trends: New perspectives on the roles of adaptation and incidental effect. *Science, 221*, 387–389.

Vrba, E. S. (1984). What is species selection? *Systematic Zoology, 33*, 318–328.

Vrba, E. S. (1989). Levels of selection and sorting with special reference to the species. *Oxford Surveys in Evolutionary Biology, 6*, 111–168.

Vrba, E. S., & Eldredge, N. (1984). Individuals, hierarchies and processes: Towards a more complete evolutionary theory. *Paleobiology, 10*, 146–171.

Vrba, E. S., & Gould, S. J. (1986). The hierarchical expansion of sorting and selection: Sorting and selection cannot be equated. *Paleobiology, 12*, 217–228.

Waddington, C. H. (1940). *Organizers and genes*. Cambridge University Press.

Wade, M. (1978). A critical review of the models of group selection. *Quarterly Review of Biology, 53*, 101–114.

Wallace, A. F. C. (1972). Paradigmatic processes in culture change. *American Anthropologist, 74*, 467–478.

Walsh, D. M. (2015). *Organisms, agency, and evolution*. Cambridge University Press.

Watkins, T. (1987). Kharabeh Shattani: An Halaf culture exposure in northern Iraq. In J. L. Huot (Ed.), *Préhistoire de la Mésopotamie* (pp. 221–230). Éditions du Centre National de la Recherche Scientifique.

Weber, M. (1904). *The Protestant ethic and the spirit of capitalism*. Scribner.

Wendorf, F. (1968). Site 117: A Nubian final Paleolithic graveyard near Jebel Sahaba, Sudan. In F. Wendorf (Ed.), *The prehistory of Nubia* (Vol. II, pp. 954–995). Southern Methodist University Press.

White, L. A. (1959). *The evolution of culture*. McGraw-Hill.

White, L. A. (1975). *The concept of cultural systems: A key to understanding tribes and nations*. Columbia University Press.

Whiting, J. W. M. (1964). Comment on Naroll "On Ethnic Unit Classification". *Current Anthropology, 5*, 305–306.

Wills, W. H. (1988). *Early prehistoric agriculture in the American southwest*. SAR Press.

Williams, G. C. (1966). *Adaptation and natural selection: A critique of some current evolutionary thought*. Princeton University Press.

Wilson, E. O. (1975). *Sociobiology: The new synthesis*. Harvard University Press.

Wilson, D. S., & Sober, E. (1994). Re-introducing group selection to the human behavioral sciences. *Behavioral and Brain Sciences, 17*, 585–654.

Wolf, E. R. (1982). *Europe and the people without history*. University of California Press.

Wrangham, R. W. (2018). Two types of aggression in human evolution. *Proc. Natl. Acad. Sci. USA, 115*(2), 245–253.

Wrangham, R. W., & Peterson, D. (1996). *Demonic males: Apes and the origins of human violence*. Houghton Mifflin Harcourt.

Yellen, J. (1985). Bushmen. *Science, 85*, 41–48.

Yellen, J. (1990). The transformation of the Kalahari !Kung. *Scientific American, 262*, 96–105.

Yoffee, N. (1991). Review of: C.K. Maisels, "The Emergence of Civilization: From Hunting-Gathering to Agriculture, Cities and the State in the Ancient Near East." *Times Literary Supplement*, January 25.

Yoffee, N. (2005). *Myths of the archaic state: Evolution of the earliest cities, states, and civilizations*. Cambridge University Press.

Yoffee, N., & Cowgill, G. (Eds.). (1988). *The collapse of ancient states and civilizations*. University of Arizona Press.

Zahavi, A., & Zahavi, A. (1997). *The handicap principle: A missing piece of the darwinian puzzle*. Oxford University Press.

Zeder, M. A. (2009). The Neolithic macro-(r)evolution: Macroevolutionary theory and the study of culture change. *Journal of Archaeological Research, 17*, 1–63.

Zeder, M. A. (2012). The broad spectrum revolution at 40: Resource diversity, intensification, and an alternative to optimal foraging xxplanations. *Journal of Anthropological Archaeology, 31*, 241–264.

Zeder, M. A. (2016). Domestication as a model system for niche construction theory. *Evolutionary Ecology, 30*, 325–348.

The manufacturer's authorised representative in the EU is Springer Nature Customer Service Centre GmbH, Europaplatz 3, 69115 Heidelberg, Germany. If you have any concerns regarding our products, please contact ProductSafety@springernature.com

Printed and bound by CPI Group (UK) Ltd, Croydon, CR0 4YY
30/09/2026
02225496-0002